THE INSECTS AND ARACHNIDS OF CANADA

PART 9

The Sac Spiders of Canada and Alaska

Araneae: Clubionidae and Anyphaenidae

Charles D. Dondale
and
James H. Redner

Biosystematics Research Institute
Ottawa, Ontario

Research Branch
Agriculture Canada

Publication 1724   1982

QL
458.42
.C4
D66
1982 / 49,624

Catalogue No. A 42-42/1982-9E   Canada: $8.95
ISBN 0-660-11146-2   Other countries: $10.75

Price subject to change without notice

The Insects and Arachnids of Canada

Part 1. Collecting, Preparing, and Preserving Insects, Mites, and Spiders, compiled by J. E. H. Martin, Biosystematics Research Institute, Ottawa, 1977.

Part 2. The Bark Beetles of Canada and Alaska (Coleoptera: Scolytidae), by D. E. Bright, Jr., Biosystematics Research Institute, Ottawa, 1976.

Part 3. The Aradidae of Canada (Hemiptera: Aradidae), by R. Matsuda, Biosystematics Research Institute, Ottawa, 1977.

Part 4. The Anthocoridae of Canada and Alaska (Heteroptera: Anthocoridae), by L. A. Kelton, Biosystematics Research Institute, Ottawa, 1978.

Part 5. The Crab Spiders of Canada and Alaska (Araneae: Philodromidae and Thomisidae), by C. D. Dondale and J. H. Redner, Biosystematics Research Institute, Ottawa, 1978.

Part 6. The Mosquitoes of Canada (Diptera: Culicidae), by D. M. Wood, P. T. Dang, and R. A. Ellis, Biosystematics Research Institute, Ottawa, 1979.

Partie 7. Genera des Trichoptères du Canada et des États adjacents, par F. Schmid, Institut de recherches biosystématiques, Ottawa, 1980.

Part 8. The Plant Bugs of the Prairie Provinces of Canada (Heteroptera: Miridae), by L. A. Kelton, Biosystematics Research Institute, Ottawa, 1980.

Contents

Acknowledgments

This work is based mainly on the spider collection in the Canadian National Collection of Insects, Arachnids, and Nematodes at Ottawa. Other notable collections were made available on loan to the authors by Dr. Brian Ainscough of the British Columbia Provincial Museum, Dr. David Barr of the Royal Ontario Museum, Mr. Donald Buckle of Saskatoon, Dr. H. W. Levi of the Museum of Comparative Zoology at Harvard University, Dr. Norman Platnick of the American Museum of Natural History in New York, Dr. R. Leech of Edmonton, Dr. Geoffrey Scudder of the University of British Columbia, and Dr. Barry Wright of the Nova Scotia Museum of Science. The authors also wish to express indebtedness to colleagues at the Biosystematics Research Institute and to Dr. B. J. Kaston for reviewing the manuscript, and to Patricia Loshak for her excellent editing.

Introduction

Part 5 in this series provides a key to spider families, a detailed treatment of the crab spiders (families Philodromidae and Thomisidae) as represented in Canada, and a glossary of anatomical terms (Dondale and Redner 1978). The present part deals with two additional families of the hunting spiders in Canada, namely, Clubionidae and Anyphaenidae, which together are called sac spiders.

The sac spiders constitute a group of two-clawed hunting spiders with a world fauna of perhaps 1500 species. They have been treated as a single family by many workers, but are treated separately here in view of the qualitative differences in tracheation and in the setae forming the claw tufts (see Comments under both families).

The term ''sac spiders'' derives from the construction by many of these spiders of flattened tubular retreats of dense white silk; the sacs are usually made in rolled leaves or under bark. Many of the members of this group are nocturnal, spending the daylight hours in the sac, whereas others are active in daylight or darkness and live mainly in the dimness between layers of plant litter in forests, bogs, or swamps. They are swift runners, and take their prey by suddenly moving upon and seizing it with their stout toothed chelicerae.

The sac spiders that live among plants are collected best by sweep nets or beating trays. The ground dwellers may be caught in pitfall traps or by searches beneath logs or stones. Preservation is by immersion in 75% ethyl or isopropyl alcohol in neoprene-stoppered homeopathic vials. The reader is referred to Part 5 of this series (Dondale and Redner 1978) for techniques pertaining to specimen examination. One modification in the present work is the illustration of spermathecae in ventral view, i.e., as seen through the wall of the epigynum, which is immersed in clove oil.

Starting with this contribution, we have decided to provide bilingual keys for all issues of this series. We hope that this will make the series more useful to our readers.

Anatomy

The sac spiders have elongate cylindrical bodies and rather stout legs (Figs. 1, 2, 6–8, 10–13, 244, 319, 332). The two body divisions, cephalothorax (*ceph*) and abdomen (*abd*), are joined by a slender pedicel. The cephalothorax is covered dorsally by a shieldlike carapace (*car*), which bears the eyes and the dorsal groove (*gr*), and is covered ventrally by a flat plate, the sternum (*st*), and the lower lip, or labium (*lab*). The principal mouthparts and the legs project to the front or sides from the membrane joining the edges of carapace and sternum.

The eyes are in four pairs arranged in two transverse rows close to the anterior margin of the carapace. They are designated as anterior medians (*ame*), anterior laterals (*ale*), posterior medians (*pme*), and posterior laterals (*ple*). Either row (viewed dorsally) may be straight, procurved, or recurved according to species or genus. Spacing within the row may also be of taxonomic importance.

The principal mouthparts are the paired pincerlike chelicerae (*chel*) and palpi. Each chelicera comprises a large basal segment and a distal fang; the fang lies, when at rest, in a groove, the margins of which are armed with small teeth. The palpi (Fig. 5) lie immediately posterior to the chelicerae; their basal segments, the coxae (*cx*), have expanded lobes (*pcxl*), which form the sides of the preoral cavity and contain glands that pour out digestive fluids over the prey. The other segments are trochanter (*tro*), femur (*fem*), patella (*pat*), tibia (*tib*), tarsus (*tar*), and a small clawlike pretarsus (*ptar*). The tarsus, tibia, and to a lesser extent the patella of the palpus of adult males (Figs. 14, 17, 18, 21) are highly modified to form the copulatory organ, the principal parts of which are a dorsal hollowed cymbium (*cym*) and the genital bulb. The genital bulb consists primarily of a convex well-sclerotized tegulum (*teg*), within which can be discerned part of the winding seminal duct, and an intromittent organ, the embolus (*e*). The embolus usually rests on a plate or membranous area called the conductor (*con*). The palpal tibia bears a stout retrolateral apophysis (*ra*), much used in classifying spiders, and more rarely a ventral (*va*) or dorsal apophysis (*da*).

The legs of sac spiders are in four pairs of approximately equal length and thickness, and are numbered I–IV from the anterior end (Fig. 8). The segments from base to tip are the coxa, trochanter, femur, patella, tibia, tarsus (subdivided into basitarsus (*btar*) and distitarsus (*dtar*)), and pretarsus (Fig. 6). The pretarsus bears two claws (*cl*) that are often hidden in a dense claw tuft (*clt*), and a scopula (*scop*) (Fig. 3). Some segments typically bear macrosetae (*mset*) (Fig. 7), the arrangement of which is often useful in classification.

The abdomen of sac spiders is elongate and cylindrical, and may be covered dorsally by an inconspicuous scutum (Fig. 10). The position of the heart is often indicated externally by a pigmented heart mark (*hm*) (Fig. 1); small muscle attachment points may be visible. Ventrally the abdomen is traversed by the genital groove (*gg*) (Fig. 2), in which lies the opening of the internal genitalia to the outside; the female's opening is protected by a well-sclerotized plate, the epigynum (*epig*), which also bears the paired copulatory openings (*co*) (Fig. 15). The copulatory openings lead inward through the copulatory tubes (*ct*) (Fig. 16) to the spermathecae (*spt*), where semen is stored until the eggs are laid. When the

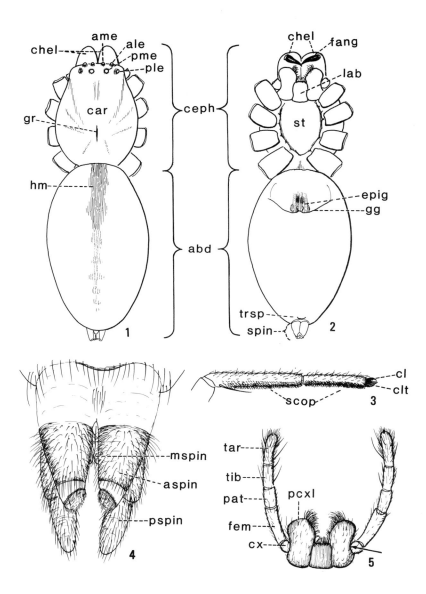

Figs. 1−5. Structures of Clubionidae. 1, Body of *Clubiona riparia*, dorsal view; 2, Body of *Clubiona riparia*, ventral view; 3, Left tarsus I of *Clubiona riparia*, prolateral view; 4, Spinnerets of *Cheiracanthium* sp., ventral view; 5, Palpi of *Clubiona riparia* female, ventral view. *abd*, abdomen; *ale*, anterior lateral eyes; *ame*, anterior median eyes; *aspin*, anterior spinnerets; *car*, carapace; *ceph*, cephalothorax; *chel*, chelicerae; *cl*, claw; *clt*, claw tuft; *cx*, coxa; *epig*, epigynum; *fem*, femur; *gg*, genital groove; *gr*, dorsal groove; *hm*, heart mark; *lab*, labium; *mspin*, median spinnerets; *pat*, patella; *pcxl*, palp-coxal lobes; *ple*, posterior lateral eyes; *pme*, posterior median eyes; *pspin*, posterior spinnerets; *scop*, scopula; *spin*, spinnerets; *st*, sternum; *tar*, tarsus; *tib*, tibia; *trsp*, tracheal spiracle.

eggs are deposited, semen is released to them via the fertilization tubes (*ft*) (Fig. 20).

Respiration is by both book lungs and tracheae. The book lungs open through a pair of slits at the lateral ends of the genital groove, and the tracheae open through a common tracheal spiracle (*trsp*) located either immediately anterior to the spinnerets (*spin*) as in the Clubionidae (Fig. 2) or farther forward on the venter as in the Anyphaenidae (Figs. 311, 315).

The spinnerets are in three pairs and form a compact cluster at the posterior end of the abdomen; these are the anterior (*aspin*), median (*mspin*), and posterior (*pspin*) spinnerets (Fig. 4).

Sac spiders are defined as those spiders in which the chelicerae close toward the midline, the eyes are arranged in two transverse rows, the legs are prograde and bear two tarsal claws, the anterior spinnerets are close together and not more heavily sclerotized than the posterior spinnerets, and the palp-coxal lobes lack a depression on the ventral surface.

Measurements given in this work include total body length, measured from the anterior margin of the carapace to the anal tubercle, to the nearest 0.05 mm; carapace length, measured from anterior to posterior ends of the carapace along the middorsal line; and carapace width, measured at the point of greatest width in dorsal aspect. Size is given as the range for fewer than 10 specimens, and as the mean and standard deviation for 10 or more.

Family Clubionidae Wagner

Spiders of the family Clubionidae have compact sparsely covered bodies that are colored in various hues of yellow, orange, or brown; some forms have iridescent scales or contrasting abdominal patterns. The body is carried close to the substratum on moderately long, strong legs. Representatives of most species are secretive and rapid of movement, and to collect them is difficult. A sweeping net or beating tray is best for those that dwell in grass or shrubs, whereas representatives of *Agroeca* spp., *Castianeira* spp., *Phrurotimpus* spp., and *Scotinella* spp. can be taken in pitfall traps set into the ground. Some of the spiders in *Castianeira*, *Phrurotimpus*, and *Scotinella* may associate with and even resemble certain ants.

Many members of this family construct flat tubular sacs, either open at the ends or closed, in rolled leaves, folded blades of grass, under loose bark, or under objects on the ground. Spiders belonging to *Clubiona* spp., *Clubionoides* spp.,

---

Figs. 6–8. Structures of Clubionidae. 6, Body of *Clubiona canadensis*, dorsal view; 7, Body of *Clubionoides excepta*, dorsal view; 8, Body of *Cheiracanthium* sp. *btar*, basitarsus; *cx*, coxa; *dtar*, distitarsus; *fem*, femur; *mset*, macrosetae; *pat*, patella; *ptar*, pretarsus; *tar*, tarsus; *tib*, tibia; *tro*, trochanter.

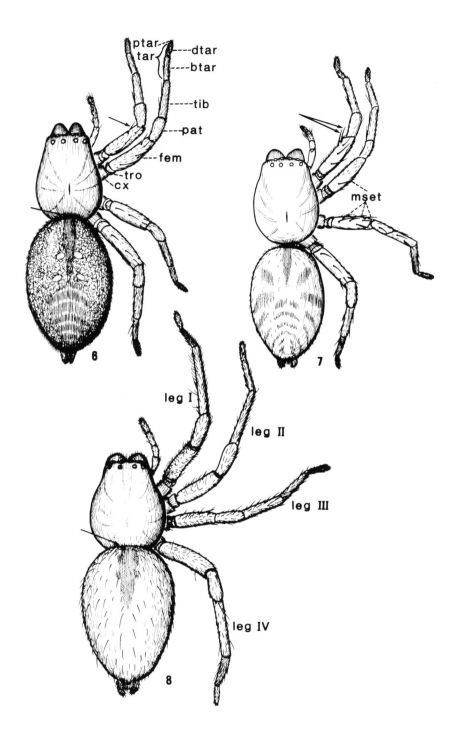

and *Cheiracanthium* spp. are in general nocturnal, and both mating and oviposition are associated with the sac. Individuals of species in the other genera, however, may be seen moving about in daylight; their eggs, which are covered with shiny, papery cases, are stuck to the underside of stones or similar objects and then abandoned. Biological information is fragmentary except for *Cheiracanthium inclusum* (Hentz), which was thoroughly studied in the laboratory by Peck and Whitcomb (1970), and *C. mildei* L. Koch, which was studied in both laboratory and field in Israel (Mansour et al., 1980*a*, *b*).

Description. Total length 1.75−10.10 mm. Carapace (Figs. 1, 6−8, 10−13, 244) ovoid in dorsal view, distinctly longer than wide, widest at level of coxae II or of coxae II and III, highest at (or anterior to) dorsal groove; dorsal groove short, shallow, usually distinct, sometimes located on posterior declivity; carapace colored in various hues of yellow, orange, or brown, occasionally marked with black, with sparse coat of short pale usually recumbent setae. Eyes small, usually uniform in size, placed in two transverse rows that may be straight, recurved, or procurved, the posterior row usually slightly longer than the anterior; posterior medians sometimes angular or ovoid in outline. Chelicerae rather long, slender, or stout, sometimes with longitudinal ridges in males; promargin of fang furrow with two to seven small teeth, and retromargin with two to four small teeth. Palp-coxal lobes usually longer than wide, without oblique depressions on ventral surfaces, sometimes constricted at middle on lateral margin. Legs prograde, yellow, orange, or brown, usually without dark rings or longitudinal bands, long and rather stout (sometimes relatively longer in males than in females), often with dense claw tufts composed of slender packed setae and with dense scopulae; trochanters with (Fig. 9) or without notches at distal ends on ventral side; tibia and basitarsus I with macrosetae on ventral surface only; tarsi with two claws. Abdomen (Figs. 1, 2) yellow, brown, or dull red, elongate-ovoid in dorsal view, rounded dorsally from side to side, sometimes bearing dorsal scutum (Fig. 10), with short semierect setae and sometimes with cluster of long erect curved setae at anterior end (Fig. 6); anterior spinnerets close together, not more heavily sclerotized than other spinnerets (Fig. 4); tracheal spiracle (Fig. 2) located immediately anteriad of anterior spinnerets. Male palpal tibia (e.g., Figs. 14, 17, 18, 21) with retrolateral apophysis variable among species, occasionally with ventral and/or dorsal apophyses; cymbium sometimes with basal spur (Fig. 21). Tegulum (Figs. 14, 17) smooth, convex, usually with apophysis at distal end; embolus usually arising prolaterodistally on tegulum, more rarely on retrolateral margin. Female epigynal plate (e.g., Figs. 15, 39, 42, 183, 224, 232, 241, 263) convex, flat, or concave, round to elongate, well-sclerotized, with copulatory openings usually distinct. Spermathecae (e.g., Figs. 16, 28, 180, 185, 227, 235, 242, 264) variable among species, each often in two connected parts of different shape, with or without spermathecal organ.

Comments. In general, spiders belonging to the family Clubionidae most resemble those of the Gnaphosidae and Anyphaenidae, which comprise hunting spiders of approximately the same size and color range and in which the leg tarsi are also two-clawed. Representatives of Clubionidae differ from those of Gnaphosidae by having the anterior spinnerets close together and not more heavily

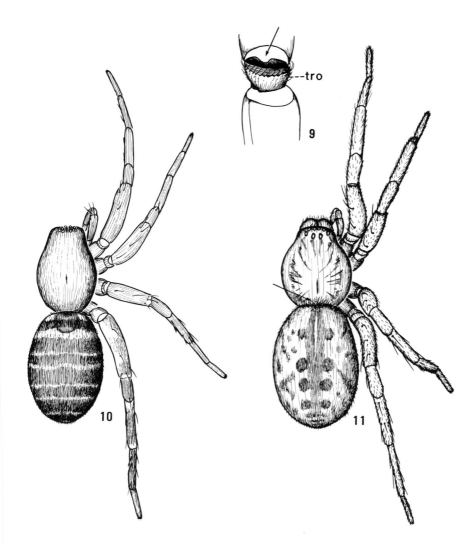

Figs. 9–11. Structures of Clubionidae. 9, Trochanter IV of *Castianeira longipalpa*, ventral view; 10, Body of *Castianeira longipalpa*, dorsal view; 11, Body of *Agroeca ornata*, dorsal view. *tro*, trochanter.

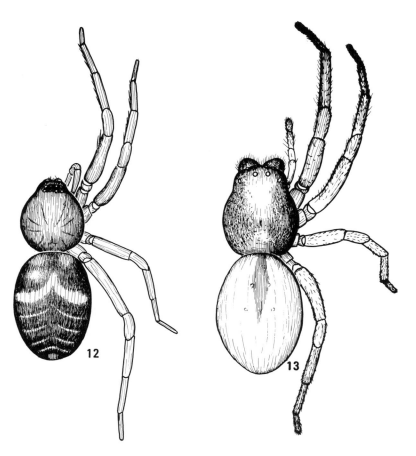

Figs. 12, 13. Bodies of Clubionidae, dorsal view. 12, *Scotinella pugnata*; 13, *Trachelas tranquillus*.

sclerotized than the posterior spinnerets, and by lacking oblique depressions on the ventral surfaces of the palp-coxal lobes. Clubionids also tend to have less-flattened abdomens and longer and less-stout legs than gnaphosids. In clubionids the tracheal spiracle is situated immediately anteriad of the anterior spinnerets rather than farther forward as in representatives of the Anyphaenidae. The tracheae are restricted to the abdomen rather than extending into the cephalothorax. In addition, many representatives of the Clubionidae have dense brushlike claw tufts, whereas anyphaenids have thin tufts consisting of only a few setae (Platnick 1974; Platnick and Lau 1975).

The Clubionidae comprise a world fauna in the order of 120 genera and 1280 species. Of these, about 22 genera and about 180 species occur in North America. Eight genera and 66 species are represented or are assumed to be represented in Canada and Alaska.

Key to genera of Clubionidae

1. Palp-coxal lobes constricted at middle along lateral margins (Fig. 5) 2
    Palp-coxal lobes straight or convex along lateral margins 4
2(1). Abdomen with dense cluster of long curved erect setae at anterior end (e.g., Fig. 6). Cymbium of male palpus without basal spur (Fig. 25). 3
    Abdomen without dense cluster of long curved erect setae at anterior end (Fig. 8). Cymbium of male palpus with strong basal spur (Figs. 17, 21) . . . .
    . . . . *Cheiracanthium* **C.L. Koch** (p. 17)
3(2). Femur I with one prolateral macroseta (Fig. 6) . . . *Clubiona* **Latreille** (p. 22)
    Femur I with two or three prolateral macrosetae (Fig. 7) . . . .
    . . . . *Clubionoides* **Edwards** (p. 97)
4(1). Posterior row of eyes distinctly procurved in dorsal view (Figs. 10, 11). Trochanter IV with distinct notch in distal margin on ventral side (Fig. 9). Basitarsus I with two or three pairs of ventral macrosetae 5
    Posterior row of eyes straight or recurved (Figs. 12, 13, 244). Trochanter IV without notch in distal margin on ventral side. Basitarsus I with four pairs of ventral macrosetae or with none 6
5(4). Basitarsus I with two pairs of ventral macrosetae. Abdomen without dense cluster of long curved erect setae at anterior end (Fig. 10) . . . .
    . . . . *Castianeira* **Keyserling** (p. 99)
    Basitarsus I with three pairs of ventral macrosetae. Abdomen with dense cluster of long curved erect setae at anterior end (Fig. 11) . . . .
    . . . . *Agroeca* **Westring** (p. 118)
6(4). Legs with macrosetae. Posterior row of eyes essentially straight in dorsal view (Figs. 12, 244). Femur of male palpus with ventral prominence (Figs. 240, 265) 7
    Legs without macrosetae. Posterior row of eyes distinctly recurved in dorsal view (Fig. 13). Femur of male palpus without ventral prominence . . . .
    . . . . *Trachelas* **L. Koch** (p.123)
7(6). Femur I with dorsal macroseta (Fig. 245). Ventral prominence of male palpal femur smoothly rounded (Fig. 240) and retrolateral apophysis of male palpal tibia with one process (Fig. 240). Copulatory openings of female seen as conspicuous cavities located midlength or farther anteriad on the epigynal plate (e.g., Figs. 241, 247). Each spermatheca in two parts, one part large and curled, the other small, ovoid or angular (e.g., Figs. 242, 246) . . . . .
    . . . . *Phrurotimpus* **Chamberlin & Ivie** (p.130)
    Femur I without dorsal macroseta (Fig. 12). Ventral prominence of male palpal femur hooked (e.g., Figs. 265, 296) and retrolateral apophysis of male palpal tibia with two processes (e.g., Figs. 262, 265). Copulatory openings of female inconspicuous, usually located at margins of paired depressions in anterior part of epigynal plate (e.g., Figs. 263, 290). Each spermatheca of single part (e.g., Figs. 264, 267, 272) . . . . *Scotinella* **Banks** (p.140)

Clé des genres de Clubionidæ

1.   Lobes coxo-palpaux rétrécis au milieu le long des marges latérales (fig. 5) . . 2

       Lobes coxo-palpaux droits ou convexes le long des marges latérales 4

2(1).   Abdomen avec touffe dense de longues soies courbées et dressées à l'extrémité antérieure (p. ex., fig. 6). Cymbium du palpe mâle sans éperon basal (fig. 25) 3

       Abdomen sans touffe dense de longues soies courbées et dressées à l'extrémité antérieure (fig. 8). Cymbium du palpe mâle avec éperon basal robuste (fig. 17 et 21) . . . . *Cheiracanthium* **C.L. Koch** (p. 17)

3(2).   Fémur I avec une macroseta prolatérale (fig. 6) . . *Clubiona* **Latreille** (p. 22)

       Fémur I avec deux ou trois macrosetæ prolatérales (fig. 7) . . . .
       . . . . *Clubionoides* **Edwards** (p. 97)

4(1).   Rangée postérieure d'yeux distinctement courbée antérieurement en vue dorsale (fig. 10 et 11). Marge distale du trochanter IV distinctement échancrée sur la face ventrale (fig. 9). Basitarse I avec deux ou trois paires de macrosetæ ventrales 5

       Rangée postérieure d'yeux droite ou recourbée (fig. 12, 13 et 244). Marge distale du trochanter IV non échancrée sur la face ventrale. Basitarse I avec quatre paires de macrosetæ ventrales ou sans aucune 6

5(4).   Basitarse I avec deux paires de macrosetæ ventrales. Abdomen sans touffe dense de longues soies courbées et dressées à l'extrémité antérieure (fig. 10) . . .
       . . . . *Castianeira* **Keyserling** (p. 99)

       Basitarse I avec trois paires de macrosetæ ventrales. Abdomen avec touffe dense de longues soies courbées et dressées à l'extrémité antérieure (fig. 11) . . .
       . . . . *Agrœca* **Westring** (p. 118)

6(4).   Pattes avec macrosetæ. Rangée postérieure d'yeux essentiellement droite en vue dorsale (fig. 12 et 244). Fémur du palpe mâle avec protubérance ventrale (fig. 240 et 265) 7

       Pattes sans macrosetæ. Rangée postérieure d'yeux distinctement recourbée en vue dorsale (fig. 13). Fémur du palpe mâle sans protubérance ventrale . . .
       . . . . *Trachelas* **L. Koch** (p. 123)

7(6).   Fémur I avec macroseta dorsale (fig. 245). Protubérance ventrale du fémur palpal mâle légèrement arrondie (fig. 240) et apophyse rétrolatérale du tibia palpal mâle avec un processus (fig. 240). Orifices copulatoires de la femelle ayant l'aspect de cavités apparentes situées au milieu de la plaque épigynale ou plus antérieurement (p. ex., fig. 241 et 247). Chaque spermathèque divisée en deux parties, l'une large et enroulée, et l'autre petite, ovoïde ou angulaire (p. ex., fig. 242 et 246) . . . . *Phrurotimpus* **Chamberlin & Ivie** (p. 130)

       Fémur I sans macroseta dorsale (fig. 12). Protubérance ventrale du fémur palpal mâle crochue (p. ex., fig. 265 et 296) et apophyse rétrolatérale du tibia palpal mâle avec deux processus (p. ex., fig. 262 et 265). Ouvertures copulatoires de la femelle peu visibles, généralement situées en marge des dépressions paires à la partie antérieure de la plaque épigynale (p. ex., fig. 263 et 290). Chaque spermathèque constituée d'une seule partie (p. ex., fig. 264, 267 et 272) . . . . *Scotinella* **Banks** (p. 140)

16

Genus *Cheiracanthium* C.L. Koch

These spiders are swift nocturnal hunters on plant foliage or on buildings. Their dense claw tufts, which are composed of many flattened setae packed in a tight bundle, presumably promote surefootedness on slippery and precipitous surfaces. The body colors are various hues of yellow, orange, or brown, though, as Peck and Whitcomb (1970) showed, ingested food may impart different colors to the partly translucent cephalothorax and abdomen.

Description. Total length 5.75−6.75 mm. Carapace (Fig. 8) yellow to orange, approximately ovoid in dorsal view, highest just anterior to dorsal groove, clothed with thin coat of short recumbent setae; dorsal groove shallow, pale, inconspicuous. Eyes nearly uniform in size, arranged in two transverse rows; anterior row slightly recurved; posterior row straight, slightly longer than anterior; eyes of posterior row uniformly spaced. Chelicerae dark orange or orange brown, protruding anteriad, with two teeth on promargin of fang furrow and two or three on retromargin. Palp-coxal lobes longer than wide, constricted at middle on lateral margin as in Fig. 5. Legs yellow, long, moderately stout, with dense claw tufts and moderately dense scopulae; basitarsus I with unpaired ventral macroseta at distal end of segment; trochanter IV with distinct ventral notch at distal end; femur and tibia II of male with one or more unusually stout macrosetae. Abdomen dull yellow or pale green, elongate-oval (Fig. 8), rather pointed posteriad, without pattern or scuta and without cluster of long erect setae at anterior end. Tibia of male palpus (Figs. 14, 17, 18, 21) with retrolateral apophysis, sometimes with dorsal and/or ventral apophyses; cymbium with strong spur at base. Tegulum with stout apophysis; embolus long, fine, arising on retrolateral side of tegulum, nearly encircling tegulum (Figs. 14, 18). Epigynum of female (Figs. 15, 19) with flat or concave plate; copulatory openings situated at lateral margins of plate. Copulatory tubes extending anteriad, then posteriad; spermathecae small, nearly round, well-separated, situated near genital groove (Fig. 16).

Comments. Characters by which specimens of *Cheiracanthium* spp. can be separated from those of related genera such as *Clubiona* and *Clubionoides* are the inconspicuous dorsal groove (Fig. 8), the presence of only two teeth on the promargin of the cheliceral fang furrow, the presence of an unpaired ventral macroseta near the tip of basitarsus I, the modified macrosetae on femur and tibia II of the male (used in mating), the lack of a cluster of long curved erect setae at the anterior end of the abdomen (Fig. 8), the cymbial spur and long fine encircling embolus of the male palpus (Figs. 14, 18), and the small well-separated spermathecae of the female (Fig. 16).

The genus *Cheiracanthium* includes a world fauna of about 150 species. Only two are represented in North America (Edwards 1958); both occur, with restricted ranges, in Canada.

Map 1. Collection localities of *Cheiracanthium* ...

Clubiona inclusa Hentz, 1847:451, fig. 18 (pl. 23).
Clubiona subflava Blackwall, 1862:426.

10–13, 17, 202; Peck 1975:204.

spermathecae small, nearly round, well-separated (fig. 16).

epigynum.

of the female's resting cell, with the pair locked together by the modified

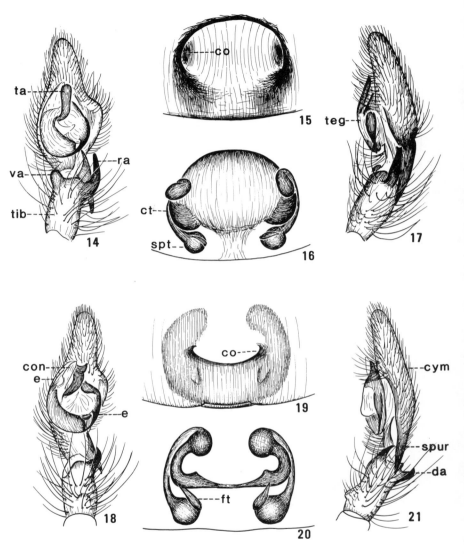

Figs. 14–21. Genitalia of *Cheiracanthium* spp. 14–17, *C. inclusum*. 14, Palpus of male, ventral view; 15, Epigynum; 16, Spermathecae, dorsal view; 17, Palpus of male, retrolateral view. 18–21, *C. mildei*. 18, Palpus of male, ventral view; 19, Epigynum; 20, Spermathecae, dorsal view; 21, Palpus of male, retrolateral view. *co*, copulatory opening; *con*, conductor; *ct*, copulatory tube; *cym*, cymbium; *da*, dorsal apophysis; *e*, embolus; *ft*, fertilization tube; *ra*, retrolateral apophysis; *spt*, spermatheca; *ta*, tegular apophysis; *teg*, tegulum; *tib*, tibia; *va*, ventral apophysis.

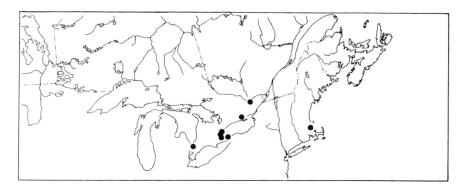

Map 2. Collection localities of *Cheiracanthium mildei*.

Cheiracanthium mildei L. Koch

Figs. 8, 18−21; Map 2

Cheiracanthium mildei L. Koch, 1864:144; Edwards 1958:371, figs. 7−9, 14, 16, 203.
Clubiona pallens Pavesi, 1864:109.

Male. Total length approximately 6.75 mm; carapace 3.12−3.24 mm long, 2.37−2.43 mm wide (three specimens measured). Carapace orange yellow. Chelicerae dark orange. Legs yellow. Abdomen pale yellow to dark yellow, with paler indistinct heart mark. Tibia of palpus with short slender tapered retrolateral apophysis and similar dorsal apophysis (Figs. 18, 21). Cymbium with basal spur extending between tibial apophyses. Tegular apophysis concave at tip; embolus long, fine, with tip lying on membranous conductor at distal end of bulb (Fig. 18).

Female. Total length approximately 6.75 mm; carapace 6.04−7.28 mm long, 2.54−3.36 mm wide (three specimens measured). General structure and color (Fig. 8) essentially as in male. Epigynum without excavation (Fig. 19); plate traversed by slightly procurved furrow, with shallow depression anterior to furrow, and with low transverse swelling posterior to it. Copulatory tubes rather stout, extending anteriad, then posteriad; spermathecae small, nearly spherical, well-separated, located short distance anterior to genital groove (Fig. 20).

Comments. Males of *C. mildei* are distinguished from those of *C. inclusum* by the presence of a dorsal apophysis on the palpal tibia and by the possession of a concave rather than a truncate tegular apophysis. Females of *mildei* posses a flat epigynal plate with a transverse furrow rather than a deep excavation.

Range. California to Alabama, northward to Illinois, southern Ontario, and Massachusetts; Europe and North Africa.

21

Genus *Clubiona* Latreille

...with one prolateral macroseta; trochanter IV with shallow notch in distal margin on ventral side. Abdomen (Fig. 6) elongate-

...apophysis at prolaterodistal angle, tibia without ventral apophysis, with stout retrolateral apophysis (often with two or more teeth), and occasionally with dorsal apophysis. Tegulum (e.g., Figs. 30, 38, 45, 100, 140, 152, 166, 174) elongate, convex, with apophysis at prolaterodistal angle near base of embolus; embolus arising at or near tip of tegular apophysis, more rarely near middle or retrolateral side of tegulum, its terminal part often arched around distal end of tegulum and extending toward base of tegulum in shallow groove; conductor not a separate piece (except in the *puchero* group). Epigynum of female (e.g., Figs. 23, 27, 39

... length on the possession of the obsidereal tone furrow, the absence of an unpaired ventral macrosets near the tip of basitarsus I, the lack of modified ... on femur and tibia II of the male, the possession of a cluster of long ...

5(4).   Retrolateral apophysis on palpal tibia with two processes that lie close together (e.g., Figs. 51, 59, 63, 81, 95) .... *abboti* **group** (p. 37)
Retrolateral apophysis on palpal tibia with single process, usually with prominence on ventral margin and/or excavation on dorsal margin (Figs. 106, 113) .... *obesa* **group** (part) (p. 62)

6(2).   Ventral part of retrolateral tibial apophysis harpoon-shaped (Figs. 149, 154, 161, 165). Conductor prominent (Figs. 147, 152, 158, 162) ....
.... *reclusa* **group** (p. 80)
Ventral part of retrolateral tibial apophysis not harpoon-shaped (e.g., Figs. 142, 169, 177). Conductor grovelike (Fig. 106) or not developed (Fig. 166) .
....7

7(6).   Embolus arising at, or basal to, middle of tegulum, arching along wall of alveolus, with base exposed (ventral view) and tip coiled (Figs. 166, 170)
.... *lutescens* **group** (p. 89)
Embolus arising at prolaterodistal angle of tegulum, not arching along wall of alveolus, with base hidden (ventral view) and tip straight or curved, not coiled (Figs. 140, 143, 174) 8

8(7).   Embolus stout, distinctly curved, with tip free (Fig. 174) ....
.... *pallidula* **group** (p. 94)
Embolus slender, nearly straight, with tip lying on part of tegulum (Figs. 140, 143) .... *obesa* **group** (part) (p. 62)

9(1).   Epigynal plate traversed by numerous fine curved ridges and grooves (e.g., Figs. 42, 53, 148, 159), and both copulatory openings and copulatory tubes located elsewhere than at midline 10
Epigynal plate without fine transverse ridges or grooves (e.g., Figs. 23, 39, 107, 167, 175), except *pygmaea* (Fig. 35) 11

10(9). Copulatory openings appearing as distinct rounded cavities (sometimes conjoined at midline) (e.g., Figs. 42, 61, 69, 87) .... *abboti* **group** (p. 37)
Copulatory openings slitlike, located in posterior margin of epigynal plate (e.g., Figs. 153, 163) .... *reclusa* **group** (p. 80)

11(9). Copulatory openings rounded, ovoid, or pocketlike, never hidden in posterior margin of epigynal plate (e.g., Figs. 23, 27, 100, 115, 127, 167, 175) ..
....12
Copulatory openings small, slitlike, hidden in posterior margin of epigynal plate (Fig. 39) .... *maritima* **group** (p. 34)

12(11). Copulatory tubes parallel, located at midline (Figs. 28, 32, 36), except *C. moesta*, in which spermathecae are sometimes covered dorsally by large plate (Fig. 24) .... *trivialis* **group** (p. 26)
Copulatory tubes not parallel, not located at midline (e.g., Figs. 100, 120, 132, 168, 176) 13

13(12). Copulatory tubes more slender than spermathecae (e.g., Figs. 105, 120, 138, 176) 14
Copulatory tubes as broad as spermathecae (Figs. 168, 172) ....
.... *lutescens* **group** (p. 89)

14(13). Copulatory openings pocketlike (Fig. 175) and both parts of spermathecae elongate (Fig. 176) .... *pallidula* **group** (p. 94)
Copulatory openings not pocketlike (e.g., Figs. 100, 115, 135, 141) or if so, then spermathecae with one or both parts round or elliptical (Figs. 123, 128) ..
.... *obesa* **group** (p. 62)

Clé de groupes d'espèces de *Clubiona*

(Femelles de *C. levii* inconnues)

1. Mâle 2
   Femelle 9
2(1). Partie terminale de l'embolus se prolongeant par la base le long d'un conducteur superficiel en forme de gouttière (p. ex., fig. 30, 38, 45 et 106) 3
   Partie terminale de l'embolus située à l'extrémité distale de la tégule, sans prolongement par la base (p. ex., fig. 140, 152, 166 et 174) 6
3(2). Apophyse rétrolatérale du tibia palpal avec un processus large et plat, indistinctement angulaire ou denté (fig. 25, 29, 33 et 37) . . . .
   . . . . **groupe *trivialis*** (p. 26)
   Apophyse rétrolatérale du tibia palpal avec un ou deux processus, mais toujours angulaires, dentés, excavés ou crochus (p. ex., fig. 38, 48, 106 et 117) . .
   4
4(3). Apophyse tégulaire calcariforme (fig. 38 et 41). Tibia palpal avec apophyse dorsale trapue (fig. 41) . . . . **groupe *maritima*** (p. 34)
   Apophyse tégulaire non calcariforme (p. ex., fig. 45, 114 et 147). Tibia palpal sans apophyse dorsale (p. ex., fig. 51 et 63) 5
5(4). Apophyse rétrolatérale du tibia palpal avec deux processus rapprochés (p. ex., fig. 51, 59, 63, 81 et 95) . . . . **groupe *abboti*** (p. 37)
   Apophyse rétrolatérale du tibia palpal avec un seul processus, généralement avec protubérance sur la marge ventrale et (ou) excavation sur la marge dorsale (fig. 106 et 113) . . . . **groupe *obesa*** (partie) (p. 62)
6(2). Partie ventrale de l'apophyse tibiale rétrolatérale sagittée (fig. 149, 154, 161 et 165). Conducteur apparent (fig. 147, 152, 158 et 162) . . . .
   . . . . **groupe *reclusa*** (p. 80)
   Partie ventrale de l'apophyse tibiale rétrolatérale non sagittée (p. ex., fig. 142, 169 et 177). Conducteur en forme de gouttière (fig. 106) ou non développé (fig. 166) 7
7(6). Embolus sortant du milieu de la tégule ou plus près de sa base, formant un arc le long de la paroi de l'alvéole, avec base exposée (vue ventrale) et extrémité enroulée (fig. 166 et 170) . . . . **groupe *lutescens*** (p. 89)
   Embolus sortant de l'angle prolatérodistal de la tégule, sans arcure le long de la paroi de l'alvéole, avec base masquée (vue ventrale) et extrémité droite ou courbée, non enroulée (fig. 140, 143 et 174) 8
8(7). Embolus trapu, distinctement courbé, avec extrémité libre (fig. 174) . . . .
   . . . . **groupe *pallidula*** (p. 94)
   Embolus étroit, presque droit, avec extrémité reposant sur une partie de la tégule (fig. 140 et 143) . . . . **groupe *obesa*** (partie) (p. 62)
9(1). Plaque épigynale traversée de nombreux fins sillons et arêtes courbés (p. ex., fig. 42, 53, 148 et 159), ouvertures et tubes copulatoires situés ailleurs que sur la ligne médiane 10
   Plaque épigynale non traversée de fins sillons ou arêtes (p. ex., fig. 23, 39, 107, 167 et 175), sauf *pygmæa* (fig. 35) 11
10(9). Ouvertures copulatoires apparaissant sous forme de cavités distinctement arrondies (parfois soudées à la ligne médiane) (p. ex., fig. 42, 61, 69 et 87) . . . . **groupe *abboti*** (p. 37)
   Ouvertures copulatoires en forme de fente, situées à la marge postérieure de la plaque épigynale (p. ex., fig. 153 et 163) . . . . . . **groupe *reclusa*** (p. 80)

The *trivialis* group

... one-half length of tegulum. Epigynum of female (Figs 22, 27, 31, 37, ...) ... with plate projecting in blunt point posteriad of ... genital groove, with ... openings united at midline or separated, usually small, usually looking ... ... or arched laterad, parallel and close together at midline, spermathecae ... large plate.

Comments. — The broad flat unadorned retrolateral apophysis on the male ... tibia, the parallel, closely-spaced copulatory tubes ... and/or the large plate ... distinguish the specimens of the *trivialis* group ... *trivialis* group. ...

Key to species of the *trivialis* group

Clé des espèces du groupe *trivialis*

6(5). Ouvertures copulatoires complètement soudées (fig. 27) ....
.... *trivialis* **C.L. Koch** (p. 30)
Ouvertures copulatoires partiellement soudées ou légèrement séparées (fig. 31 et
35) 7

7(6). Tubes copulatoires se prolongeant vers l'avant jusqu'à l'extrémité antérieure des
spermathèques, mesurant environ la moitié de la largeur des spermathèques
(fig. 32) .... *quebecana* **Dondale & Redner** (p. 32)
Tubes copulatoires se prolongeant vers l'avant au niveau des parties postérieures
des spermathèques, mesurant nettement moins de la moitié de la largeur des
spermathèques (fig. 36) .... *pygmæa* **Banks** (p. 34)

Clubiona moesta Banks

Figs. 22–25; Map 3

Clubiona pusilla Emerton, 1890:181, figs. 5–5*b* (pl.5). Name *pusilla* preoccupied in genus *Clubiona*.

Clubiona moesta Banks, 1896:64; Edwards 1958:393, figs. 99, 100, 130, 144, 228.

Clubiona emertoni Petrunkevitch, 1911:460.

Clubiona orinoma Chamberlin, 1919*b*:255, fig. 4 (pl. 14).

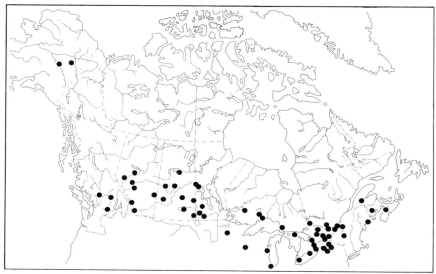

Map 3. Collection localities of *Clubiona moesta*.

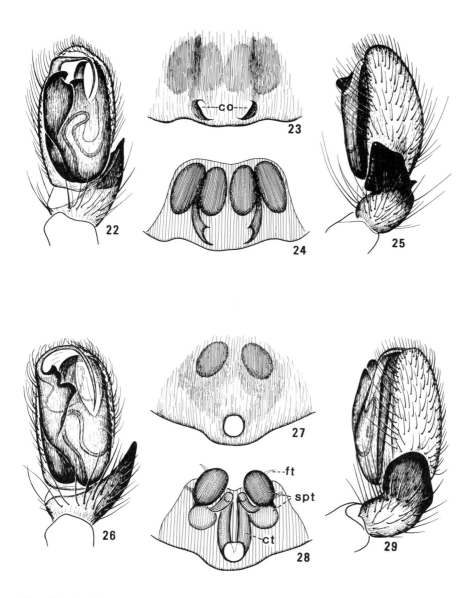

Figs. 22−29. Genitalia of *Clubiona* spp. 22−25, *C. moesta*. 22, Palpus of male, ventral view; 23, Epigynum; 24, Spermathecae, ventral view; 25, Palpus of male, retrolateral view. 26−29, *C. trivialis*. 26, Palpus of male, ventral view; 27, Epigynum; 28, Spermathecae, ventral view; 29, Palpus of male, retrolateral view. *co*, copulatory openings; *ct*, copulatory tube; *ft*, fertilization tube; *spt*, spermatheca.

long, 1.42 ± 0.15 mm wide (20 specimens measured). Carapace orange yellow to
orange brown. Chelicerae orange brown to dull red, protruding, each with
approximately five long procurate setae on anterior surface and with long shallow

conductor less than one-third length of tegulum (Fig. 2x).

Female. Total length approximately 4.60 mm. Carapace 2.12 ± 0.18 mm

transversely, and covered dorsally by large plate (Fig. 2v).

Range. Alaska to Nova Scotia, southward to Colorado and to Long Island,
New York.

sac was found in June.

Figs. 26—29, Map 4

Male. Total length approximately 3.55 mm. Carapace 1.61 ± 0.13 mm
long, 1.18 ± 0.08 mm wide (20 specimens measured). Carapace yellow orange.
Chelicerae orange brown, orange red hairs on anterior surface. Legs orange to
yellow. Abdomen dull red to dull yellow, rarely near black, with thin scutum

Map. Collection localities of *Chelifer cancroides*.

Justar (Fig. 26)

Description. Total length approximately 4.10 mm. Carapace 1.61 ± 0.12 mm long, 1.17 ± 0.09 mm wide (20 specimens measured). General structure and

by large plate (Fig. 28)

Comments. Specimens of *C. trivialis* closely resemble those of the other

conjoined copulatory openings of females.

Europe and Asia.

Biology. Specimens of *C. trivialis* are common inhabitants of caves fin

Clubiona quebecana Dondale & Redner

Figs. 30–33; Map 5

Clubiona quebecana Dondale & Redner, 1976:1157, figs. 1–4.

Male.   Total length approximately 4.00 mm; carapace 1.49–1.82 mm long, 1.06–1.36 mm wide (four specimens measured). Carapace yellow orange, slightly suffused with green or gray, darkest anteriorly. Chelicerae orange brown, with anterior surfaces rugose. Legs orange yellow. Abdomen yellow, finely speckled with dark red, with indistinct heart mark. Tibia of palpus with broad pointed retrolateral apophysis (Fig. 33). Embolus arising at distal end of tegular apophysis, curved around distal end of tegulum, terminating along conductor slightly basad of midpoint of tegulum (Fig. 30). Tegular apophysis smoothly rounded at tip (Fig. 30).

Female.   Total length approximately 4.25 mm. Carapace 1.57–1.81 mm long, 1.10–1.27 mm wide (three specimens measured). General structure and color essentially as in male but paler (venter lacking red tone). Epigynum with copulatory openings moderately large, close together (Fig. 31). Copulatory tubes approximately one-half as wide as spermathecae, close together, parallel, extending anteriad to level of anterior parts of spermathecae; each spermatheca in two rounded parts, the anterior part slightly larger than the posterior (Fig. 32).

Comments.   Specimens of *C. quebecana* most resemble those of *C. pygmaea* but can be distinguished from the latter by having a longer embolus and more rounded tegular apophysis in the male and by having longer and less slender copulatory tubes in the female.

Range.   Wisconsin to southern Quebec and Massachusetts.

Biology.   The available specimens of *C. quebecana* were collected from the trunks and larger branches of deciduous trees such as oaks. Mature males were collected in April, May, and August, and mature females in March, May, and August.

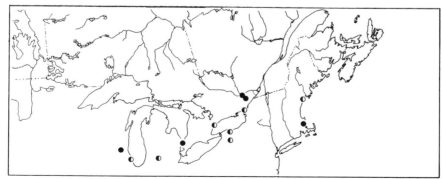

Map 5. Collection localities of *Clubiona quebecana* (●) and *C. pygmaea* (◖).

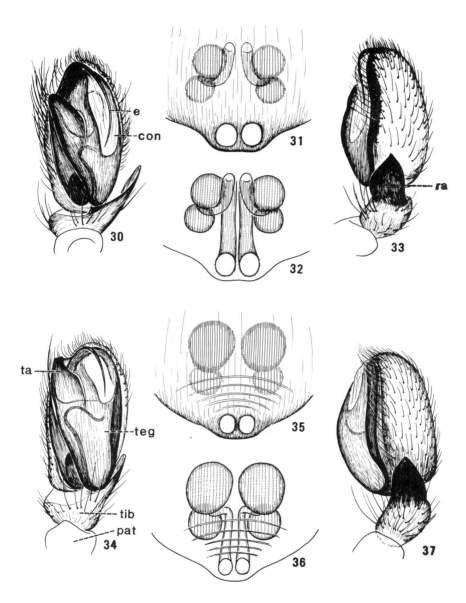

Figs. 30–37. Genitalia of *Clubiona* spp. 30–33, *C. quebecana*. 30, Palpus of male, ventral view; 31, Epigynum; 32, Spermathecae, ventral view; 33, Palpus of male, retrolateral view. 34–37, *C. pygmaea*. 34, Palpus of male, ventral view; 35, Epigynum; 36, Spermathecae, ventral view; 37, Palpus of male, retrolateral view. *con*, conductor; *e*, embolus; *pat*, patella; *ra*, retrolateral apophysis; *ta*, tegular apophysis; *teg*, tegulum; *tib*, tibia.

Clubiona pygmaea Banks

Figs. 34—37; Map 5

Clubiona minuta Emerton, 1890:181, figs. 11—11*b* (pl. 5). Name *minuta* preoccupied in genus *Clubiona*.
Clubiona pygmaea Banks, 1892:21, fig. 64 (pl. 1).
Clubiona lenta Banks, 1892:21, fig. 66 (pl. 1).
Clubiona minutissima Petrunkevitch, 1911:461.

Male. Total length approximately 3.25 mm. Carapace 1.50 mm long, 1.03 mm wide (one specimen measured). Carapace yellow orange, sometimes nearly black in eye area. Chelicerae dark orange brown. Legs pale orange. Abdomen dull red to off-white. Tibia of palpus with somewhat pointed retrolateral apophysis (Fig. 37). Tegulum convex, with tegular apophysis bearing small hook at tip (Fig. 34); embolus slender, arched around distal end of tegulum, with its tip lying on membranous conductor and extending one-fifth to one-third length of tegulum (Fig. 34).

Female. Total length approximately 3.80 mm. Carapace 1.56, 1.59 mm long, 1.04, 1.16 mm wide (two specimens measured). General structure and color essentially as in male. Epigynum with plate produced in blunt prominence posteriad over genital groove; copulatory openings narrowly separated at midline (Fig. 35). Copulatory tubes less than one-half as wide as spermathecae, straight, close together, parallel, extending anteriad to level of posterior parts of spermathecae; spermathecae in two parts, with anterior part rounded, larger than and directly anteriad of posterior part (Fig. 36).

Comments. Specimens of *C. pygmaea* most resemble those of *C. quebecana*, but can be distinguished by the shorter embolus and hooked tegular apophysis in males and by the shorter and more slender copulatory tubes in females.

Range. Colorado to southern Ontario and Maine, southward to Texas and Florida.

Biology. Recorded habitats for *C. pygmaea* include tall grass in marshes and low deciduous shrubs. Adults of both sexes have been collected in June and July.

The *maritima* group

Description. Total length 6.05—6.50 mm. Palpal tibia of male with short slender retrolateral apophysis bearing strong ventral process and two small teeth on prolateral margin, and with large strong dorsal apophysis (Figs. 38, 41). Tegulum with prominent spurlike apophysis that extends distad as far as tip of cymbium (Fig. 38); embolus arising in deep cavity between tegular apophysis and tegular wall, slender, tapered, arched around distal end of tegulum, extending along membranous conductor to base of tegulum (Fig. 38). Epigynum of female with broad plate having median prominence and posterior excavation; copulatory openings small, slitlike, well-separated, located obscurely in posterior rim of plate (Fig. 39). Copulatory tubes long, looped laterad; spermathecae elongate (Fig.40), lying in pocket between ventral wall of epigynal plate and copulatory tubes.

Comments. The long spurlike tegular apophysis, the dorsal apophysis on the male palpal tibia, and the obscure slitlike copulatory openings distinguish specimens of the *maritima* group from those of other groups. One species occurs in Canada.

Clubiona maritima L. Koch

Figs. 38−41; Map 6

Clubiona maritima L. Koch, 1866:310, fig. 198 (pl. 12); Edwards 1958:432, figs. 131−133, 139, 180, 214.
Clubiona tibialis Emerton, 1890:180, figs. 3−3*b* (pl. 5).
Clubiona transversa Bryant, 1936:97, fig. 8 (pl. 3).

Male. Total length approximately 6.00 mm. Carapace 2.51−3.02 mm long, 1.71−2.15 mm wide (six specimens measured). Carapace orange or yellow orange. Chelicerae dark orange brown, with long depression on anteromesal surface. Legs orange yellow. Abdomen yellow red, often becoming uniform gray in preservative. Patella of palpus with short broad apophysis near prolateral margin. Tibia of palpus with short slender retrolateral apophysis bearing strong ventral process and two small prolateral teeth (Figs. 38, 41) and with broad dorsal apophysis bearing broad tooth and sinuous process (Fig. 41). Tegular apophysis long, spurlike, arising midlength tegulum; embolus arising in deep cavity between tegular apophysis and tegular wall, slender, tapered, arched around distal end of tegulum, extending along groovelike conductor to base of tegulum (Fig. 38).

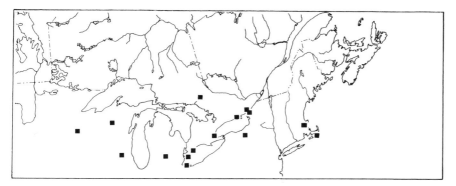

Map 6. Collection localities of *Clubiona maritima*.

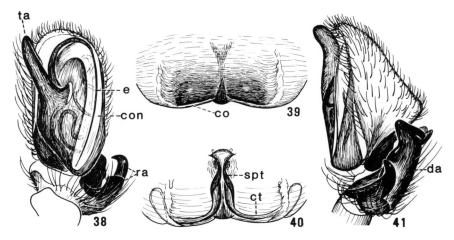

Figs. 38–41. Genitalia of *Clubiona maritima*. 38, Palpus of male, ventral view; 39, Epigynum; 40, Spermathecae, ventral view; 41, Palpus of male, retrolateral view. *co*, copulatory opening; *con*, conductor; *ct*, copulatory tube; *da*, dorsal apophysis; *e*, embolus; *ra*, retrolateral apophysis; *spt*, spermatheca; *ta*, tegular apophysis.

Female.   Total length approximately 6.00 mm. Carapace 2.40–3.24 mm long, 1.58–2.24 mm wide (seven specimens measured). General structure and color essentially as in male but chelicerae lacking depression along anteromesal surface. Epigynum with broad plate having broad excavation in posterior margin and having low median prominence that divides and extends laterad along posterior margin; copulatory openings small, slitlike, inconspicuous, located in posterior margin of epigynal plate (Fig. 39). Copulatory tubes long, extending anteriad along lateral margins of epigynum, then looping and extending mesad to midline, finally extending anteriad together; spermathecae elongate, bent, lying within pocket between ventral wall of epigynal plate and copulatory tubes (Fig. 40).

Comments.   *C. maritima* is the only species known to date in the *maritima* group, and its distinguishing characters are those of the group.

Range.   Minnesota to Ontario and Massachusetts, southward to Texas, Florida, and the West Indies.

Biology.   Specimens of *C. maritima* have been collected under loose tree bark, under stones, under fallen tree branches on the ground, from the leaf sheaths of cattails, and from tall grass in a marsh. The egg sac is made within a folded grass blade. Mature males have been taken in May, July, and October, and mature females from April to July.

The *abboti* group

Description. Total length 2.50−6.10 mm. Retrolateral apophysis of male palpal tibia (e.g., Figs. 48, 51, 63, 95) with two processes; a stout ventral process, which is either truncated or excavated at tip and which is expanded at some point along ventral margin, and a tapered dorsal process, which may be either shorter or longer than ventral process. Tegulum elongate, convex, with prominent apophysis arising near its middle; apophysis twisted distad and usually deeply hollowed on retrolateral margin, concealing base of embolus in ventral view (e.g., Figs. 45, 60, 79); embolus (Figs. 45, 60, 79) long, tapered, with distal part lying along shallow groovelike conductor near retrolateral margin of tegulum. Epigynum of female (e.g., Figs. 42, 61, 69, 87) with plate traversed by numerous fine grooves and ridges; copulatory openings usually conspicuous, cavitylike, close together near posterior margin of plate. Copulatory tubes (e.g., Figs. 43, 44, 62, 70, 88) slender, sinuous, or smoothly arched, sometimes with small coil, extending anterolaterad; spermathecae situated close together anteriorly, in two parts that are approximately equal in size, one lying anterior or anterolateral to the other.

Comments. The two-part retrolateral apophysis of males and the usually distinct and rounded copulatory openings and the ridged and grooved epigynal plate of females are together diagnostic for the *abboti* group. Thirteen species occur or are assumed to occur in Canada.

Key to species of the *abboti* group

Clé des espèces du groupe *abboti*

Clubiona abboti L. Koch

Figs. 42–45, 47; Map 7

Clubiona abbotii L. Koch, 1866:303, fig. 193 (pl. 12); Edwards 1958:417, figs. 42, 43, 83, 181, 182, 236.
Clubiona rubra Keyserling, 1887:436, fig. 12 (pl. 6).
Clubiona bufonis Chamberlin, 1925:220.
Clubiona abbotoides Chamberlin & Ivie, 1946:10, figs. 13, 14.

Male. Total length approximately 3.75 mm; carapace 1.79 ± 0.13 mm long, 1.23 ± 0.11 mm wide (20 specimens measured). Carapace orange to orange yellow, in some specimens paler anterior to dorsal groove and with several pale lines radiating from dorsal groove. Chelicerae dark orange brown, without ridges along anterior or lateral surfaces. Legs pale orange. Abdomen yellow orange to dull red, with inconspicuous scutum covering anterior two-thirds of dorsum. Patella of palpus with blunt prolaterodistal apophysis. Tibia of palpus with retrolateral apophysis with two processes; ventral process broad throughout its length and with narrow excavation at tip; dorsal process shorter, more slender, pointed (Fig. 47). Tegular apophysis large, U-shaped, with arms of U approximately equal in length and thickness (Fig. 45); embolus broad at base, tapered, arched around distal end of tegulum, with distal part slender, extending basad along conductor approximately one-half length of tegulum (Fig. 45).

Female. Total length approximately 4.50 mm; carapace 1.88 ± 0.19 mm long, 1.31 ± 0.15 mm wide (20 specimens measured). General structure and color essentially as in male but abdomen lacking dorsal scutum. Epigynum with large plate; plate with shallow depression anteriad, with numerous fine transverse grooves and ridges, and with posterior bilobed prominence; copulatory openings small, angular, close together, separated from posterior margin of epigynal plate by approximately their length (Fig. 42). Copulatory tubes slender, slightly curved,

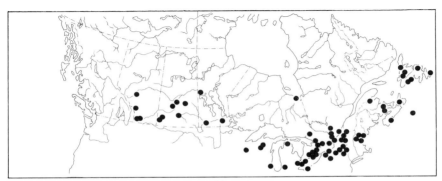

Map 7. Collection localities of *Clubiona abboti*.

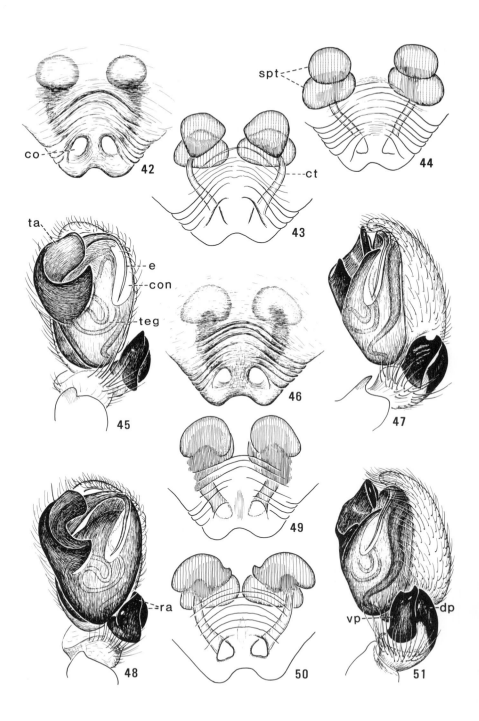

extending anterolaterad; spermathecae in two parts, with anterior part round or subtriangular in outline, not extending as far laterad as posterior part (Figs. 43, 44).

Comments. Males of *C. abboti* closely resemble those of *C. bishopi* but differ in having a narrower excavation at the tip of the ventral process of the retrolateral tibial apophysis. Females of *abboti* resemble those of *gertschi* in having the copulatory openings small, angular, and well-separated from the posterior margin of the epigynal plate, but can be distinguished from the latter by the spermathecae, the anterior parts of which extend less far laterad than in female *bishopi*.

Range. Alberta to Newfoundland, southward to Texas and Florida.

Biology. Specimens of *C. abboti* have been collected from shrubs and herbs of many kinds, among roots and under stones and plant litter in meadows, beaches, and deciduous woods, and on calcareous and sphagnum bogs. Both sexes have been collected from March to September, and females until October.

Clubiona bishopi Edwards

Figs. 46, 48−51; Map 8

Clubiona bishopi Edwards, 1958:413, figs. 40, 41; Dondale & Redner 1976:1162, figs. 21−26.

Male. Total length approximately 3.90 mm; carapace 1.78 ± 0.07 mm long, 1.28 ± 0.04 mm wide (19 specimens measured). Carapace orange to orange yellow, slightly suffused with gray or green. Chelicerae orange brown, without ridges on anterior or lateral surfaces. Legs yellow to orange. Abdomen yellow, with several curved transverse rows of orange dots on posterior half and with inconspicuous orange scutum covering anterior half of dorsum. Patella of palpus with short pointed ventral apophysis. Tibia of palpus with retrolateral apophysis with two processes; ventral process slightly swollen along ventral margin, with broad excavation at tip; dorsal process stout, pointed, slightly shorter than ventral process (Figs. 48, 51). Tegular apophysis large, U-shaped, with both arms of U approximately of same length and thickness (Fig. 48); embolus thick at base, tapered, arching around tip of tegulum, extending basad along tegulum approximately one-half length of tegulum (Fig. 48).

Figs. 42−51. Genitalia of *Clubiona* spp. 42−45, 47, *C. abboti*. 42, Epigynum; 43, 44, Spermathecae, ventral view; 45, Palpus of male, ventral view; 47, Palpus of male, retrolateral view. 46, 48−51, *C. bishopi*. 46, Epigynum; 48, Palpus of male, ventral view; 49, 50, Spermathecae, ventral view; 51, Palpus of male, retrolateral view. *co*, copulatory opening; *con*, conductor; *ct*, copulatory tube; *dp*, dorsal process; *e*, embolus; *ra*, retrolateral apophysis; *spt*, spermatheca; *ta*, tegular apophysis; *teg*, tegulum; *vp*, ventral process.

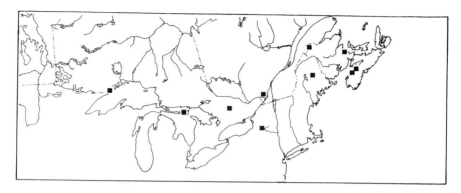

Map 8. Collection localities of *Clubiona bishopi*.

Female. Total length approximately 4.55 mm; carapace 1.93 ± 0.12 mm long, 1.32 ± 0.12 mm wide (15 specimens measured). General structure and color essentially as in male but abdomen lacking dorsal scutum. Epigynum with large plate; plate with numerous fine transverse grooves and ridges, and with bilobed prominence at posterior margin; copulatory openings small, angular, separated from posterior margin of epigynal plate by approximately their length (Fig. 46). Copulatory tubes rather slender, extending anterolaterad; spermathecae in two parts, with anterior part extending farther laterad than posterior part (Figs. 49, 50).

Comments. Individuals of *C. bishopi* closely resemble those of *C. abboti*. Males of *bishopi* have a broader excavation at the tip of the ventral process of the retrolateral tibial apophysis, and females have the anterior parts of the spermathecae extending farther laterad.

Range. Ontario to Nova Scotia, southward to North Carolina.

Biology. Individuals of *C. bishopi* have been collected from herbs and shrubs of many kinds, by pitfall traps in deciduous forest litter, and on sphagnum bogs. Both sexes have been collected from May to September.

Clubiona kastoni Gertsch

Figs. 52−55; Map 9

Clubiona kastoni Gertsch, 1941*b*:14, figs. 37−39; Edwards 1958:414, figs. 46, 47, 81, 196, 233.

Male. Total length approximately 4.10 mm; carapace 1.85 ± 0.10 mm long, 1.39 ± 0.07 mm wide (20 specimens measured). Carapace yellow or dull orange, with few paler or darker lines radiating from dorsal groove. Chelicerae dark orange, lacking ridges on prolateral and lateral surfaces. Legs yellow or pale orange. Abdomen pale orange to dull red, with inconspicuous scutum covering anterior three-fourths of dorsum. Patella of palpus with blunt prominent ventral

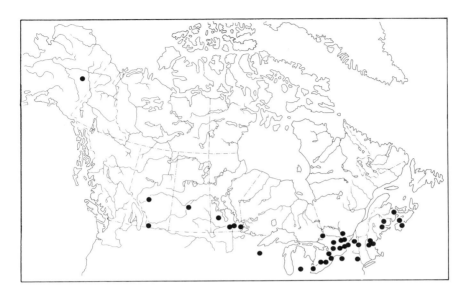

Map 9. Collection localities of *Clubiona kastoni*.

apophysis. Tibia of palpus with retrolateral apophysis with two processes; ventral process broad, truncate; dorsal process more slender, pointed, tapered near tip (Fig. 55). Tegular apophysis large, fluted at tip, deeply excavated on retrolateral margin (Fig. 52); embolus broad at base, tapered, arched around distal end of tegulum, extending basad along tegulum approximately one-half length of tegulum (Fig. 52).

Female.   Total length approximately 4.65 mm; carapace 2.02 ± 0.13 mm long, 1.41 ± 0.07 mm wide (17 specimens measured). General structure and color essentially as in male but abdomen lacking dorsal scutum. Epigynum with large plate; plate with median depression anteriad, with numerous fine transverse grooves and ridges, and with bilobed prominence at posterior margin; copulatory openings large, close together, close to posterior margin of epigynal plate (Fig. 53). Copulatory tubes rather thick, extending anterolaterad; spermathecae each in two parts, with anterior part ovoid in shape and with posterior part approximately rectangular (Fig. 54).

Comments.   Specimens of *C. kastoni* closely resemble those of *C. johnsoni*. Males of *kastoni* are distinguished from those of *johnsoni* by the lack of a dorsal tooth on the dorsal process of the retrolateral apophysis. Females of *kastoni* have the anterior parts of the spermathecae ovoid rather than round in outline.

Range.   Alaska to Nova Scotia, southward to California and to North Carolina.

Biology.   Specimens of *C. kastoni* were collected by pitfall traps in deciduous forest litter, on beaches and sand dunes, or on bogs. A few were collected on low shrubs. Both sexes have been collected from April to September.

45

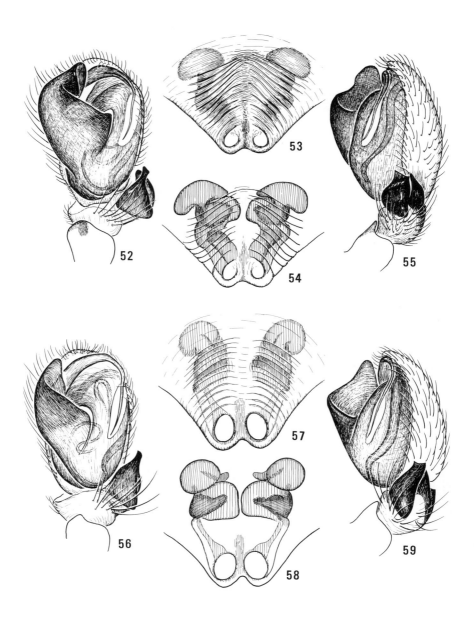

Figs. 52−59. Genitalia of *Clubiona* spp. 52−55, *C. kastoni*. 52, Palpus of male, ventral view; 53, Epigynum; 54, Spermathecae, ventral view; 55, Palpus of male, retrolateral view. 56−59, *C. johnsoni*. 56, Palpus of male, ventral view; 57, Epigynum; 58, Spermathecae, ventral view; 59, Palpus of male, retrolateral view.

Clubiona johnsoni Gertsch

Figs. 56—59; Map 10

Clubiona johnsoni Gertsch, 1941*b*:14, figs. 43—45; Edwards 1958:422, figs. 48, 49, 96, 187, 235.

Male. Total length approximately 3.25 mm; carapace 1.55 ± 0.07 mm long, 1.04 ± 0.05 mm wide (20 specimens measured). Carapace orange or orange yellow, sometimes with several paler lines radiating from dorsal groove. Chelicerae dark orange brown, without ridges along anterior or lateral surfaces. Legs pale orange. Abdomen yellow orange to dull red, with inconspicuous scutum covering anterior half to two-thirds of dorsum. Patella of palpus with blunt ventral apophysis. Tibia of palpus with retrolateral apophysis with two processes; ventral process slightly tapered toward tip, broadly excavated at tip; dorsal process slightly shorter than ventral process, with small tooth (Fig. 59). Tegular apophysis large, not U-shaped or fluted (Fig. 56); embolus broad at base, tapered, arched around distal end of tegulum, extending basad slightly more than one-half length of tegulum (Fig. 56).

Female. Total length approximately 3.80 mm; carapace 1.66 ± 0.10 mm long, 1.12 ± 0.06 mm wide (20 specimens measured). General structure and color essentially as in male but abdomen lacking dorsal scutum. Epigynum with large plate; plate depressed mesad, with bilobed prominence at posterior margin, and with numerous fine transverse grooves and ridges; copulatory openings large, round or ovoid, situated close together, close to posterior margin of epigynal plate (Fig. 57). Copulatory tubes slender, nearly straight, extending anterolaterad; spermathecae each in two parts, with anterior part rounded and posterior part approximately rectangular in outline (Fig. 58).

Comments. Specimens of *C. johnsoni* most resemble those of *C. kastoni*. Males of *johnsoni* can be distinguished by the more simple tegular apophysis, by the longer and more slender ventral process of the retrolateral apophysis on the

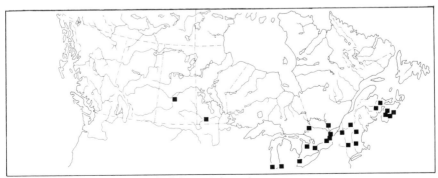

Map 10. Collection localities of *Clubiona johnsoni*.

palpal tibia, and by the presence of a small tooth on the dorsal process. Females of *johnsoni* have rounded rather than ovoid anterior parts of the spermathecae.

Range.   Saskatchewan to Nova Scotia, southward to Illinois and to Delaware.

Biology.   Specimens of *C. johnsoni* have been collected by pitfall traps in meadows, bogs, and deciduous or coniferous forests, and from shrubs and beach litter. Adults of both sexes have been collected from May to October. A female with her egg sac was collected in early August.

Clubiona angulata Dondale & Redner

Figs. 60–63; Map 11

Clubiona angulata Dondale & Redner, 1976:1164, figs. 30, 32.

Male.   Total length approximately 3.40 mm. Carapace 1.60–1.65 mm long, 1.08–1.15 mm wide (three specimens measured). Carapace pale orange yellow, slightly suffused with green. Chelicerae dark orange yellow. Legs pale orange yellow. Abdomen orange yellow, with inconspicuous orange scutum covering anterior third of dorsum. Patella of palpus with short blunt ventral apophysis. Tibia of palpus with retrolateral apophysis with two processes; ventral process untapered, broadly and shallowly excavated at tip, with both ventral and dorsal margins sinuous; dorsal process shorter, tapered, truncate at tip, with large dorsal tooth (Fig. 63). Tegular apophysis broad, rounded at tip, with angular prominence on retrolateral margin (Fig. 60); embolus broad at base, tapered, angular, extending basad approximately one-third length of tegulum (Fig. 60).

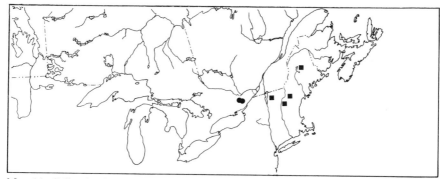

Map 11. Collection localities of *Clubiona angulata* (●) and *C. gertschi* (■).

Female. Total length approximately 3.95 mm; carapace 1.76, 1.91 mm long, 1.17, 1.28 mm wide (two specimens measured). General structure and color essentially as in male but abdomen lacking scutum. Epigynum with large dark plate (Fig. 61); plate with shallow mesal depression in posterior half, with bilobed prominence at posterior margin, and with numerous fine transverse grooves and ridges; copulatory openings small, somewhat indistinct, located at midline in posterior third of plate (Fig. 61). Copulatory tubes slender, forming small coil at posterior end, extending anterolaterad; spermathecae in two parts, with posterior part large, dark, rounded and with anterior part small, pale, flattened, located anterolaterad of posterior part (Fig. 62).

Comments. Males of *C. angulata* differ from those of other species in the *abboti* group in having an angular prominence on the retrolateral margin of the tegular apophysis and in having a truncate tip and large dorsal tooth on the dorsal process of the retrolateral apophysis of the palpal tibia. Females differ in having inconspicuous copulatory openings and closely set coiled copulatory tubes.

Range. Eastern Ontario.

Biology. An adult male of *C. angulata* was collected in a pitfall trap in early July on the wooded bank of a small river. Other mature males and females were collected in the leaf litter of a calcareous bog in early August.

Clubiona gertschi Edwards

Figs. 64–67; Map 11

Clubiona gertschi Edwards, 1958:408, figs. 50, 51, 84, 195, 232.

Male. Total length approximately 3.50 mm; carapace 1.40 mm long, 1.00 mm wide (one specimen measured). Carapace orange. Chelicerae dark orange. Legs pale yellow orange. Abdomen red orange, with few inconspicuous chevrons posteriad. Tibia of palpus with retrolateral apophysis with two processes; ventral process broadest at base, tapered, oblique at tip, smoothly curved along ventral margin; dorsal process tapered, pointed, shorter than ventral process (Fig. 67). Tegular apophysis broad, concave along retrolateral margin (Fig. 64); embolus broad at base, tapered, slightly sinuous, extending basad along tegulum approximately three-fourths length of tegulum (Fig. 64).

Female. Total length approximately 4.00 mm; carapace 1.60 mm long, 1.10 mm wide (one specimen measured). General structure and color essentially as in male. Epigynum with large plate; plate with shallow median depression, with numerous fine transverse grooves and ridges, and with bilobed prominence at posterior margin; copulatory openings large, ovoid, situated close to posterior margin of epigynal plate, separated by inconspicuous median septum (Fig. 65). Copulatory tubes rather thick, somewhat sinuous, extending anterolaterad then anteriad; each spermatheca in two parts, with anterior part flattened or angular in outline and with posterior part approximately rectangular in outline (Fig. 66).

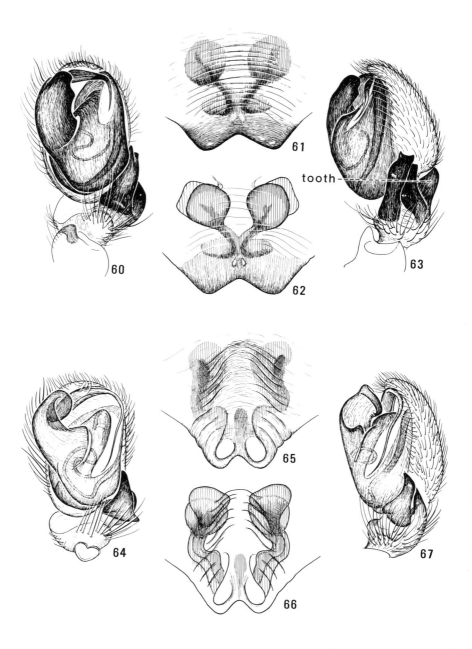

tooth-

Figs. 60−67. Genitalia of *Clubiona* spp. 60−63, *C. angulata*. 60, Palpus of male, ventral view; 61, Epigynum; 62, Spermathecae, ventral view; 63, Palpus of male, retrolateral view. 64−67, *C. gertschi*. 64, Palpus of male, ventral view; 65, Epigynum; 66, Spermathecae, ventral view; 67, Palpus of male, retrolateral view.

Comments. Males of *C. gertschi* most resemble those of *C. opeongo* but differ in having a long shallow excavation along the ventral margin of the ventral process of the retrolateral apophysis on the palpus. Females of *gertschi* differ from those of the other species in the group by having well-separated cavitylike copulatory openings situated close to the posterior margin of the epigynal plate and having the posterior parts of the spermathecae as close together as the anterior parts.

Range. Vermont, New Hampshire, and Maine.

Biology. Specimens of *C. gertschi* have been collected under stones in the higher mountains of northern New England.

Clubiona opeongo Edwards

Figs. 72−75; Map 12

Clubiona opeongo Edwards, 1958:412, figs. 86, 234; Dondale & Redner 1976:1161, figs. 16−20.

Male. Total length approximately 3.25 mm; carapace 1.44−1.80 mm long, 1.05−1.22 mm wide (eight specimens measured). Carapace yellow orange, slightly suffused with green or gray. Chelicerae yellow orange. Legs orange yellow. Abdomen orange yellow, slightly patterned with dull red, with pale orange scutum covering anterior half of dorsum. Patella of palpus with hollowed ventral apophysis. Tibia of palpus with retrolateral apophysis with two processes; ventral process with angular prominence and deep excavation on ventral margin; dorsal process pointed, tapered (Fig. 75). Tegular apophysis stout, hollowed, curved (Fig. 72); embolus broad at base, tapered, arched around distal end of tegulum, extending basad approximately three-fourths length of tegulum (Fig. 72).

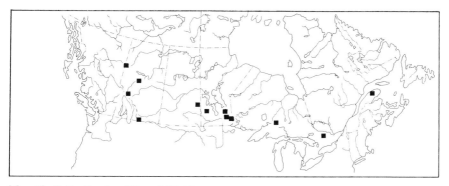

Map 12. Collection localities of *Clubiona opeongo*.

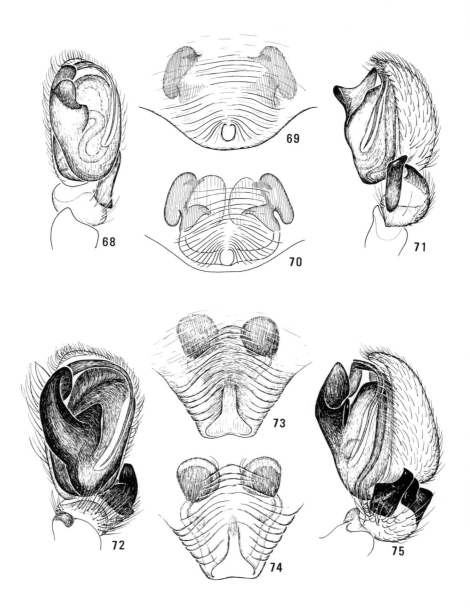

52

Female.   Total length approximately 3.75 mm; carapace 1.70 ± 0.08 mm long, 1.17 ± 0.06 mm wide (20 specimens measured). General structure and color essentially as in male but abdomen without dorsal scutum. Epigynum with large plate; plate with shallow median depression, with numerous fine transverse grooves and ridges, and with truncate or bilobed prominence at posterior margin; copulatory openings shallow, conjoined at midline (Fig. 73). Copulatory tubes somewhat sinuous, extending anterolaterad; spermathecae each in two parts, with anterior part pale, nearly round, and with posterior part darker, hollowed posteriad (Fig. 74).

Comments.   Males of *C. opeongo* most resemble those of *C. gertschi* but differ from the latter in having an angular prominence and a deep excavation along the ventral margin of the ventral process of the retrolateral tibial apophysis. Females of *opeongo* resemble those of *C. littoralis* and *C. catawba* in having conjoined copulatory openings but differ in having round anterior parts of the spermathecae.

Range.   Alberta to Quebec.

Biology.   Specimens of *C. opeongo* have been collected from black spruce or from sphagnum bogs and plant litter in spruce forests. Adults of both sexes have been taken from June to September.

Clubiona catawba Gertsch

Figs. 68−71; Map 13

Clubiona catawba Gertsch, 1941*b*:10, figs. 10, 11; Edwards 1958:426, figs. 76, 77, 92, 194, 244.
Clubiona alachua Gertsch, 1941*b*:4, fig. 4.

Male.   Total length approximately 2.50 mm; carapace 1.20 mm long, 0.87 mm wide (one specimen measured). Carapace orange. Chelicerae orange brown, each with two ridges along anteromesal surface. Legs orange yellow. Abdomen yellow, with inconspicuous scutum covering anterior one-third to one-half of dorsum. Patella of palpus with blunt ventral apophysis. Tibia of palpus with retrolateral apophysis with two processes; ventral process long, slender, straight; dorsal process shorter, tapered (Fig. 71). Tegular apophysis with two rounded prominences of approximately equal size (Fig. 68); embolus slender, tapered, arched around distal end of tegulum, extending basad approximately three-fourths length of tegulum (Fig. 68).

Figs. 68−75. Genitalia of *Clubiona* spp. 68−71, *C. catawba*. 68, Palpus of male, ventral view; 69, Epigynum; 70, Spermathecae, ventral view; 71, Palpus of male, retrolateral view. 72−75, *C. opeongo*. 72, Palpus of male, ventral view; 73, Epigynum; 74, Spermathecae, ventral view; 75, Palpus of male, retrolateral view.

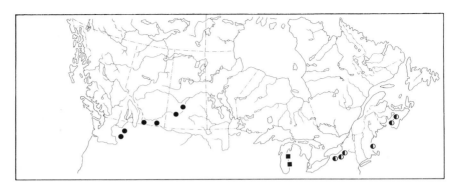

Map 13. Collection localities of *Clubiona catawba* (■), *C. mutata* (●), and *C. saltitans* (◐).

Female.   Total length approximately 3.35 mm; carapace 1.41−1.49 mm long, 0.87−0.99 mm wide (seven specimens measured). General structure and color essentially as in male but abdomen lacking dorsal scutum and chelicerae lacking ridges. Epigynum with indistinct plate; plate with shallow mesal depression, with numerous fine transverse grooves and ridges, and with broad rounded prominence at posterior margin; copulatory openings forming single small round depression, separated by less than one-half their length from posterior margin of epigynal plate (Fig. 69). Copulatory tubes rather broad, extending laterad then arched anteriad; spermathecae in two parts, with anterior part kidney-shaped, oblique, and with posterior part irregular (Fig. 70).

Comments.   Specimens of *C. catawba* are distinguished from the other members of the *abboti* group by the long straight ventral process of the retrolateral tibial apophysis and the two prominences of approximately equal size on the tegular apophysis of the male, and by the single small round copulatory opening of the female.

Range.   Texas to Florida, northward to Michigan.

Biology.   Gertsch's (1941*b*) type series of *C. alachua* was collected in March and April, and that of *C. catawba* in June. Other specimens have been collected by sweep nets in soybean fields in summer.

Clubiona mutata Gertsch

Figs. 76−79, 81; Map 13

Clubiona mutata Gertsch, 1941*b*:14, figs. 19, 20; Edwards 1958:428, figs. 64, 65, 80, 185, 246.

Male.   Total length approximately 2.80 mm; carapace 1.24−1.45 mm long, 0.82−0.98 mm wide (three specimens measured). Carapace orange.

Chelicerae dark orange or orange brown, without ridges along anterior or lateral surfaces. Legs orange yellow. Abdomen yellow orange, with inconspicuous scutum covering anterior one-third to one-half of dorsum. Patella of palpus with blunt ventral apophysis. Tibia of palpus with retrolateral apophysis with two processes; ventral process slender, with angular prominence near base on ventral margin (Fig. 81); dorsal process shorter, pointed, with angular dorsal prominence visible in both retrolateral and dorsal views (Figs. 77, 81). Tegular apophysis prominent, hollowed on retrolateral side (Fig. 79); embolus broad at base, tapered, arched around distal end of tegulum, extending basad approximately three-fourths length of tegulum (Fig. 79).

Female. Total length approximately 3.70 mm; carapace 1.45−1.65 mm long, 0.96−1.16 mm wide (six specimens measured). General structure and color essentially as in male but abdomen lacking dorsal scutum. Epigynum with large plate; plate with shallow median depression, numerous fine transverse grooves and ridges, and with bilobed prominence at posterior margin; copulatory openings large, ovoid, close together, close to posterior margin of epigynal plate (Fig. 76). Copulatory tubes rather slender, arched as far laterad as spermathecae; spermathecae each in two parts, with anterior part flattened and with posterior part approximately rectangular (Fig. 78).

Comments. *C. mutata* is defined here, as in Gertsch (1941*b*), as a western species distinguishable from eastern forms on the basis of subtle differences in the external male genitalia. The angular prominence on the dorsal process of the retrolateral tibial apophysis, together with the broad tip of the ventral process of that apophysis, distinguish the males. Females closely resemble those of *C. kiowa*, *C. pikei*, and *C. saltitans*, but tend to have shorter copulatory openings than in *kiowa* and less-round openings than in *saltitans*. *C. mutata* is currently regarded as a western species in contrast to the other three.

Range. Washington to Saskatchewan and Nebraska, southward to Arizona.

Biology. Specimens of *C. mutata* have been collected in pitfall traps from short grass prairie in Saskatchewan and Alberta, and from semi-arid grass and ponderosa pine associations in interior Oregon. Males were collected from May to July, and females from May to September.

Clubiona kiowa Gertsch

Figs. 80, 82−85; Map 14

Clubiona kiowa Gertsch, 1941*b*:12, figs. 23, 24; Edwards 1958:428, figs. 62, 63, 90, 186, 245.

Male. Total length approximately 2.90 mm; carapace 1.34 mm long, 0.84 mm wide (one specimen measured). Carapace orange. Chelicerae orange brown, without longitudinal ridges. Legs yellow or orange yellow. Abdomen yellow orange, with narrow inconspicuous scutum covering anterior one-third of

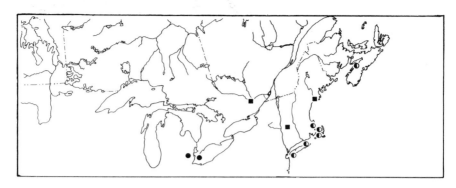

Map 14. Collection localities of *Clubiona pikei* (■), *C. littoralis* (◑), and *C. kiowa* (●).

dorsum. Patella of palpus with small prolaterodistal apophysis. Tibia of palpus with retrolateral apophysis with two processes; ventral process truncate at tip in retrolateral view, pointed in dorsal view; with large angular prominence near base on ventral margin; dorsal process shorter, pointed, with angular dorsal prominence (Figs. 84, 85). Tegular apophysis prominent, hollowed on retrolateral side, with single point at tip (Fig. 82); embolus broad at base, arched around tip of tegulum, extending basad approximately three-fourths length of tegulum (Fig. 82).

Female.   Total length approximately 3.45 mm; carapace 1.45, 1.56 mm long, 0.88, 1.03 mm wide (two specimens measured). General structure and color essentially as in male but abdomen lacking dorsal scutum. Epigynum with large plate; plate with bilobed prominence at posterior margin, with numerous transverse grooves and ridges; copulatory openings narrowly ovoid (Fig. 80). Copulatory tubes slender, slightly arched laterad; spermathecae each in two parts, with anterior part somewhat flattened, separated from opposite member of pair by distance greater than the width of one spermatheca, rather remote from copulatory openings (Fig. 83).

Comments.   Males of *C. kiowa* can be distinguished from those of the other species in the *abboti* group by having both a pointed tip (dorsal view) on the ventral process of the retrolateral tibial apophysis and an angular prominence (retrolateral view) on the dorsal process of that apophysis. Females of *kiowa* closely resemble those of *C. pikei*, *C. mutata*, and *C. saltitans* but tend to have a lesser arch in the copulatory tubes and a greater distance between spermathecae and copulatory openings.

Range.   Mexico northward to southern Ontario.

Biology.   A female of *C. kiowa* was collected in plant litter in a marsh in July.

Figs. 76–85. Genitalia of *Clubiona* spp. 76–79, 81, *C. mutata*. 76, Epigynum; 77, Tibia of male palpus, dorsal view; 78, Spermathecae, ventral view; 79, Palpus of male, ventral view; 81, Palpus of male, retrolateral view. 80, 82–85, *C. kiowa*. 80, Epigynum; 82, Palpus of male, ventral view; 83, Spermathecae, ventral view; 84, Tibia of male palpus, dorsal view; 85, Palpus of male, retrolateral view.

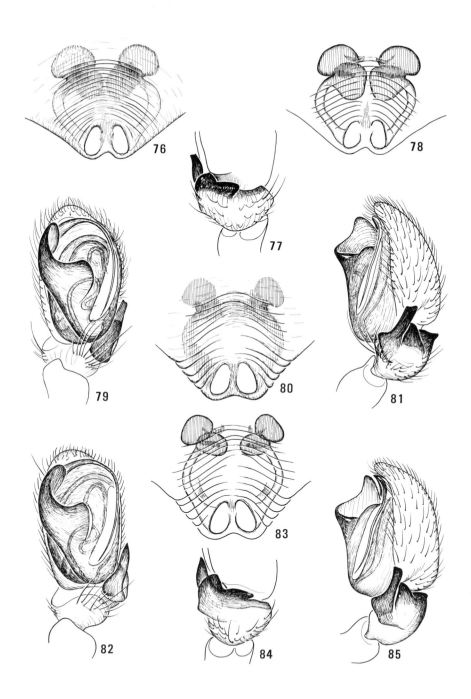

76

78

77

79

80

81

82

83

84

85

Clubiona littoralis Banks

Figs. 86–89; Map 14

Clubiona littoralis Banks, 1895:79; Edwards 1958:412, figs. 70, 71, 87, 201, 242.

Clubiona latifrons Emerton, 1913:220, figs. 12, 12a, 12b, (pl. 2). Name *latifrons* preoccupied in genus *Clubiona*.

Male. Total length approximately 5.65 mm; carapace 2.46, 2.77 mm long, 1.63, 1.64 mm wide (two specimens measured). Carapace orange, orange red, or orange brown. Chelicerae dark orange, swollen, protruding, without ridges along anterior or lateral surfaces. Legs orange yellow. Abdomen orange yellow, without dorsal scutum. Patella of palpus with blunt ventral apophysis. Tibia of palpus with retrolateral apophysis with two processes; ventral process with angular prominence on ventral margin, and with tip oblique, slightly excavated; dorsal process approximately as long as ventral process, pointed, with rounded dorsal prominence (Fig. 89). Tegular apophysis with large excavation along retrolateral margin, with distinct notch between two points of equal size at tip (Fig. 86); embolus broad at base, tapered, arched around distal end of tegulum, extending basad nearly three-fourths length of tegulum (Fig. 86).

Female. Total length approximately 6.20 mm; carapace 1.98–2.64 mm long, 1.31–1.80 mm wide (nine specimens measured). General structure and color essentially as in male. Epigynum with large plate; plate with bilobed prominence at posterior margin; copulatory openings forming large triangular depression at midline (Fig. 87). Copulatory tubes slender, extending anterolaterad; spermathecae each in two parts, with anterior part flattened, oblique (Fig. 88).

Comments. The long embolus, together with the near equality in length of the two processes of the retrolateral tibial apophysis in males, and the large triangular depression forming the copulatory openings, together with the elongate anterior parts of the spermathecae in females, distinguish specimens of *C. littoralis* from those of the other species in the *abboti* group.

Range. Florida northward to Nova Scotia.

Biology. Collections of *C. littoralis* are from salt marshes along the Atlantic coast.

Clubiona saltitans Emerton

Figs. 90–95; Map 13

Clubiona saltitans Emerton, 1919:107, figs. 14–14c; Edwards 1958:427, figs. 58, 59, 188, 248.

Figs. 86–95. Structures of *Clubiona* spp. 86–89, *C. littoralis*. 86, Palpus of male, ventral view; 87, Epigynum; 88, Spermathecae, ventral view; 89, Palpus of male, retrolateral view. 90–95, *C. saltitans*. 90, Epigynum; 91, 93, Spermathecae, ventral view; 92, Palpus of male, ventral view; 94, Chelicerae, anterior view; 95, Palpus of male, retrolateral view.

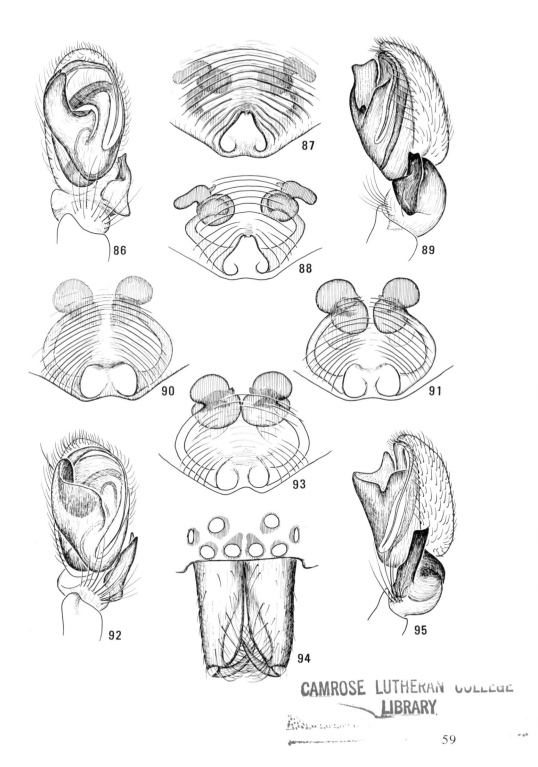

86 87 88 89 90 91 92 93 94 95

Male.   Total length approximately 3.10 mm; carapace 1.43, 1.56 mm long, 0.96, 1.07 mm wide (two specimens measured). Carapace orange. Chelicerae orange brown, each with two ridges along anteromesal surface (Fig. 94). Legs orange yellow. Abdomen yellow orange, with inconspicuous scutum covering anterior half or more of dorsum. Patella of palpus with blunt ventral apophysis. Tibia of palpus with retrolateral apophysis with two processes; ventral process with low prominence near base on ventral margin, with oblique excavated tip; dorsal process shorter, broader (Fig. 95). Tegular apophysis prominent, hollowed on retrolateral side (Fig. 92); embolus broad at base, tapered, arched around distal end of tegulum, extending basad more than three-fourths length of tegulum (Fig. 92).

Female.   Total length approximately 5.30 mm; carapace 2.13 mm long, 1.44 mm wide (one specimen measured). General structure and color essentially as in male but chelicerae lacking ridges and abdomen lacking dorsal scutum. Epigynum with large plate; plate with shallow median depression, numerous fine transverse grooves and ridges, and with bilobed prominence at posterior margin; copulatory openings large, narrowly ovoid, close together, close to posterior margin of epigynal plate (Fig. 90). Copulatory tubes arched laterad farther than spermathecae; spermathecae each in two parts, with anterior part somewhat flattened, well-separated, and with posterior part approximately rectangular (Figs. 91, 93).

Comments.   Males of *C. saltitans* most resemble those of *C. pikei* but differ in having cheliceral ridges. Females of *saltitans* closely resemble those of *C. mutata*, *C. pikei*, and *C. kiowa*, but their copulatory openings tend to be more ovoid.

Range.   Florida northward to New York and Nova Scotia.

Biology.   Kaston (1948) reports collections of *C. saltitans* from spring flood debris and from beach litter in March and June. One specimen was collected by a sweep net from soybean fields.

Clubiona pikei Gertsch

Figs. 96−99; Map 14

Clubiona pikei Gertsch, 1941*b*:10, figs. 25−27; Edwards 1958:420, figs. 56, 57, 91, 189, 249.

Male.   Total length approximately 3.05 mm; carapace 1.45−1.47 mm long, 0.91−0.99 mm wide (three specimens measured). Carapace orange yellow to pale orange brown. Chelicerae orange brown, without ridges along anterior or lateral surfaces. Legs yellow or orange yellow. Abdomen pale yellow to off-white, with inconspicuous scutum covering anterior one-third to one-half of dorsum. Patella of palpus with blunt ventral apophysis. Tibia of palpus with retrolateral apophysis with two processes; ventral process with angular prominence near base on ventral margin, with distal part straight, with tip oblique, slightly excavated;

dorsal process much shorter than ventral, pointed, extending laterad along dorsal margin of tibia (Fig. 99). Tegular apophysis prominent, hollowed on retrolateral side; embolus thick at base, tapered, arched around distal end of tegulum, extending basad approximately three-fourths length of tegulum (Fig. 96).

Female. Total length approximately 3.60 mm; carapace 1.42−1.56 mm long, 0.97−1.08 mm wide (six specimens measured). General structure and color essentially as in male, but abdomen lacking dorsal scutum. Epigynum with large plate; plate with shallow mesal depression, with numerous fine transverse grooves and ridges, and with bilobed prominence at posterior margin; copulatory openings narrowly ovoid, situated close together near posterior margin of epigynal plate (Fig. 97). Copulatory tubes slender, smoothly arched far laterad; spermathecae each in two parts, with anterior part slightly flattened and with posterior part somewhat rectangular in outline, hollowed posteriad (Fig. 98).

Comments. Males of *C. pikei* most resemble those of *C. saltitans* but differ in lacking cheliceral ridges. Females of *pikei* closely resemble those of *C. mutata*, *C. saltitans*, and *C. kiowa* but tend to have less-ovoid copulatory openings than *saltitans* and to have the spermathecae closer to the copulatory openings than *kiowa*, and a more easterly range than *mutata*.

Range. Florida northward to southern Ontario and Maine.

Biology. Specimens of *C. pikei* have been collected in pan traps among deciduous trees and shrubs. Mature males have been taken in July and mature females from July to September.

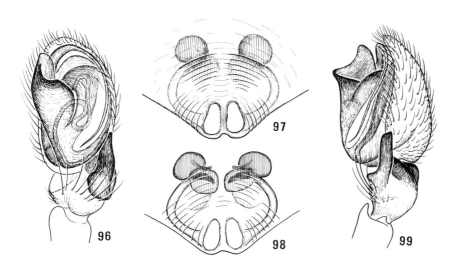

Figs. 96−99. Genitalia of *Clubiona pikei*. 96, Palpus of male, ventral view; 97, Epigynum; 98, Spermathecae, ventral view; 99, Palpus of male, retrolateral view.

The *obesa* group

Description. Total length 3.80−8.25 mm. Retrolateral apophysis on male palpal tibia broad, usually with swelling on ventral margin and with excavation on dorsal margin (e.g., Figs. 106, 117, 126). Tegulum convex, with tegular apophysis variously shaped among species; embolus arising prolaterodistally on tegulum, arched around distal end of tegulum, usually slender, usually extending basad along membranous conductor (e.g., Figs. 103, 114, 124). Epigynum of female with plate broad, smooth, convex, indented along posterior margin; copulatory openings usually conspicuous, cavitylike, located near posterolateral angles of epigynal plate (e.g., Fig. 100). Copulatory tubes slender to moderately wide, usually extending mesad or laterad, then anteriad (e.g., Figs. 100, 115, 127, 141); spermathecae each in two parts, with one or both parts round or elliptical in outline (e.g., Figs. 105, 120, 138).

Comments. Members of the *obesa* group are distinguished from those of other groups by the broad retrolateral apophysis, which usually has a swelling on the ventral margin and an excavation on the dorsal margin, by the absence of ridges and grooves on the epigynal plate, and by the presence of distinct cavitylike copulatory openings, slender non-parallel copulatory tubes, and round or elliptical spermathecae. Nine species occur in Canada or Alaska.

Key to species of the *obesa* group

(Female of *C. levii* is unknown)

1.	Male 2	
	Female 10	
2(1).	Embolus arising directly from tegulum, with base visible in ventral view (Figs. 103, 110) 3	
	Embolus arising basad of tip of tegular apophysis, with base hidden in ventral view (e.g., Figs. 114, 124, 140) 4	
3(2).	Retrolateral apophysis of male palpal tibia with dorsal excavation broad (Figs. 101, 106) .... *obesa* **Hentz** (p. 65)	
	Retrolateral apophysis with dorsal excavation narrow (Figs. 108, 113). .... *mixta* **Emerton** (p. 67)	
4(2).	Distal part of embolus extending basad along tegulum (e.g., Figs. 114, 124, 134) 5	
	Distal part of embolus erect, not extending basad along tegulum (Figs. 140, 143) 9	
5(4).	Tegular apophysis curved or hooked at tip (Figs. 114, 118, 124) 6	
	Tegular apophysis straight at tip, not hooked or curved (Figs. 129, 134) 8	
6(5).	Retrolateral tibial apophysis with two well-separated teeth; ventral tooth hooked (Fig. 117)............................. *spiralis* **Emerton** (p. 69)	
	Retrolateral tibial apophysis with one tooth, with or without ventral lobe (Figs. 121, 126) 7	
7(6).	Retrolateral tibial apophysis with simple fingerlike ventral tooth, without ventral lobe (Fig. 121) .... *mimula* **Chamberlin** (p.70)	
	Retrolateral tibial apophysis without ventral tooth, with rounded ventral lobe (Figs. 125, 126)........................ *chippewa* **Gertsch** (p.72)	

8(5). Retrolateral tibial apophysis narrower than long (Figs. 130, 131) ....
.... *furcata* **Emerton** (p. 73)
Retrolateral tibial apophysis much broader than long (Figs. 136, 137) ....
.... *praematura* **Emerton** (p. 76)
9(4). Ventral tooth on retrolateral tibial apophysis with single sharp point (retrolateral
view, Fig. 142) .... *bryantae* **Gertsch** (p. 77)
Ventral tooth on retrolateral tibial apophysis with two blunt points (retrolateral
view, Fig. 146) .... *levii* **Holm** (p. 78)
10(1). Copulatory openings ovoid, deep, located at posterolateral angles of epigynal
plate (Figs. 100, 102, 107, 109) 11
Copulatory openings not ovoid, not deep, not located at posterolateral margins of
plate (e.g., Figs 115, 127, 135) 12
11(10). Copulatory tubes folded (Figs. 104, 105). Excavation in posterior margin of
epigynal plate deep, usually with small median prominence (Figs. 100, 102)
.... *obesa* **Hentz** (p. 65)
Copulatory tubes not folded (Figs. 111, 112). Excavation in posterior margin of
epigynal plate shallow, without median prominence (Figs. 107, 109) ....
.... *mixta* **Emerton** (p. 67)
12(10). Copulatory openings transverse, close to midline (Fig. 115) ....
.... *spiralis* **Emerton** (p. 69)
Copulatory openings not transverse, removed from midline (e.g., Figs. 119, 127,
135) 13
13(12). Copulatory openings visible as oblique cavities (Fig. 127) ....
.... *furcata* **Emerton** (p. 73)
Copulatory openings not oblique (e.g., Figs. 119, 122, 135) 14
14(13). Copulatory openings distinct, angular, slightly separated, located at posterior
margin of epigynal plate (Fig. 119) ...... *mimula* **Chamberlin** (p. 70)
Copulatory openings indistinct, well-separated, located in margins of shallow
depressions (Figs. 122, 135, 141) 15
15(14). Median notch at posterior margin of epigynal plate as wide as deep (Fig. 141)
.... *bryantae* **Gertsch** (p. 77)
Median notch at posterior margin of epigynal plate much wider than deep (Figs.
122, 135) 16
16(15). Copulatory tubes extending mesad to midline, with rounded enlargement (Fig.
123) .... *chippewa* **Gertsch** (p. 72)
Copulatory tubes extending mesad then anteriad, without rounded enlargement
(Figs. 132, 133) .... *praematura* **Emerton** (p. 76)

Clé des espèces du groupe *obesa*

(Femelle de *C. levii* inconnue)

1. Mâle 2
Femelle 10
2(1). Embolus sortant directement de la tégule, avec base visible en vue ventrale
(fig. 103 et 110) 3
Embolus sortant vers la base de l'extrémité de l'apophyse tégulaire, avec base
masquée en vue ventrale (p. ex., fig. 114, 124 et 140) 4
3(2). Apophyse rétrolatérale du tibia palpal mâle avec cavité dorsale large (fig. 101 et
106) .... *obesa* **Hentz** (p. 65)
Apophyse rétrolatérale avec cavité dorsale étroite (fig. 108 et 113) ....
.... *mixta* **Emerton** (p. 67)

4(2). Partie distale de l'embolus se prolongeant vers la base le long de la tégule (p. ex., fig. 114, 124 et 134) 5

Partie distale de l'embolus dressée, ne se prolongeant pas vers la base le long de la tégule (fig. 140 et 143) 9

5(4). Apophyse tégulaire courbée ou crochue à son extrémité (fig. 114, 118 et 124) 6

Apophyse tégulaire droite à son extrémité, non crochue ou courbée (fig. 129 et 134) 8

6(5). Apophyse tibiale rétrolatérale avec deux dents bien séparées; dent ventrale crochue (fig. 117) .... *spiralis* **Emerton** (p. 69)

Apophyse tibiale rétrolatérale avec une dent, avec ou sans lobe ventral (fig. 121 et 126) 7

7(6). Apophyse tibiale rétrolatérale avec une seule dent ventrale digitiforme, sans lobe ventral (fig. 121) .... *mimula* **Chamberlin** (p. 70)

Apophyse tibiale rétrolatérale sans dent ventrale, avec lobe ventral arrondi (fig. 125 et 126)......................... *chippewa* **Gertsch** (p. 72)

8(5). Apophyse tibiale rétrolatérale plus étroite que longue (fig. 130 et 131) .... *furcata* **Emerton** (p. 73)

Apophyse tibiale rétrolatérale beaucoup plus large que longue (fig. 136 et 137) .... *prœmatura* **Emerton** (p. 76)

9(4). Dent ventrale sur l'apophyse tibiale rétrolatérale avec une seule pointe effilée (vue rétrolatérale, fig. 142) .... *bryantœ* **Gertsch** (p. 77)

Dent ventrale sur l'apophyse tibiale rétrolatérale avec deux pointes obtuses (vue rétrolatérale, fig. 146) .... *levii* **Holm** (p. 78)

10(1). Ouvertures copulatoires ovoïdes, profondes, situées aux angles postérolatéraux de la plaque épigynale (fig. 100, 102, 107 et 109)................ 11

Ouvertures copulatoires non ovoïdes, non profondes, non situées aux marges postérolatérales de la plaque (p. ex., fig. 115, 127 et 135) 12

11(10). Tubes copulatoires pliés (fig. 104 et 105). Cavité profonde sur la marge postérieure de la plaque épigynale, généralement avec petite protubérance médiane (fig. 100 et 102) .... *obesa* **Hentz** (p. 65)

Tubes copulatoires non pliés (fig. 111 et 112). Cavité superficielle sur la marge postérieure de la plaque épigynale, sans protubérance médiane (fig. 107 et 109) .... *mixta* **Emerton** (p. 67)

12(10). Ouvertures copulatoires transversales, proches de la ligne médiane (fig. 115) . ... *spiralis* **Emerton** (p. 69)

Ouvertures copulatoires non transversales, éloignées de la ligne médiane (p. ex., fig. 119, 127 et 135) 13

13(12). Ouvertures copulatoires visibles sous forme de cavités obliques (fig. 127) .... *furcata* **Emerton** (p. 73)

Ouvertures copulatoires non obliques (p. ex., fig. 119, 122 et 135) 14

14(13). Ouvertures copulatoires distinctes, angulaires, légèrement séparées, situées à la marge postérieure de la plaque épigynale (fig. 119) .... *mimula* **Chamberlin** (p. 70)

Ouvertures copulatoires indistinctes, bien séparées, situées à la marge de dépressions peu profondes (fig. 122, 135 et 141) 15

15(14). Échancrure médiane à la marge postérieure de la plaque épigynale aussi large que profonde (fig. 141) .... *bryantœ* **Gertsch** (p. 77)

Échancrure médiane à la marge postérieure de la plaque épigynale beaucoup plus large que profonde (fig. 122 et 135) 16

16(15). Tubes copulatoires se prolongeant jusqu'à la ligne médiane, avec élargissement arrondi (fig. 123) .... *chippewa* **Gertsch** (p. 72)

Tubes copulatoires se prolongeant vers la ligne médiane, puis antérieurement, sans élargissement arrondi (fig. 132 et 133) ....
.... *præmatura* **Emerton** (p. 76)

Clubiona obesa Hentz

Figs. 100–106; Map 15

Clubiona obesa Hentz, 1847:450, fig. 14 (pl. 22); Edwards 1958:398, figs. 121–123, 140, 172, 224.
Clubiona crassipalpis Keyserling, 1887:438, fig. 13.
Clubiona triloba Banks, 1907:737, fig. 19.

Male. Total length approximately 7.00 mm; carapace 2.58–3.65 mm long, 1.92–2.62 mm wide (nine specimens measured). Carapace orange. Chelicerae orange red, with low ridge along lateral margin and with shallow depression along anteromesal surface. Legs yellow orange. Abdomen dull orange red, with inconspicuous orange scutum covering anterior one-half or one-third. Femur of palpus with short blunt ventral apophysis. Tibia of palpus with retrolateral apophysis convex on ventral margin, with large excavation on dorsal margin, broad and truncate at tip (Figs. 101, 106). Tegular apophysis not well developed; embolus broad at base, arched around distal end of tegulum, extending in gentle curve approximately one-half length of tegulum along groovelike conductor (Fig. 103).

Female. Total length approximately 8.25 mm; carapace 3.44 ± 0.37 mm long, 2.50 ± 0.21 mm wide (20 specimens measured). General structure and color essentially as in male but chelicerae lacking lateral ridge and anteromesal depression, and abdomen lacking dorsal scutum. Epigynum with broad somewhat convex plate having deep posterior indentation and usually small median

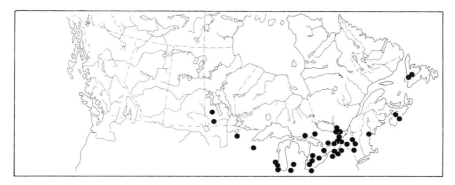

Map 15. Collection localities of *Clubiona obesa*.

prominence; copulatory openings conspicuous, cavitylike, located near posterolateral angles of epigynal plate (Figs. 100, 102). Copulatory tubes bent upon themselves; spermathecae each in two parts occupying approximately one-half width of epigynal plate, with anterior part pale, rounded, and with posterior part darker, elongate, smaller (Figs. 104, 105).

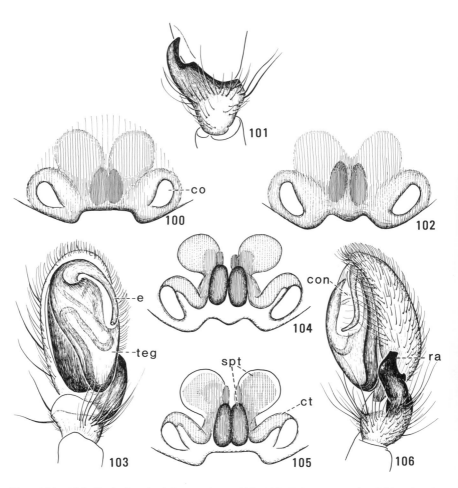

Figs. 100–106. Genitalia of *Clubiona obesa*. 100, 102, Epigynums; 101, Tibia of male palpus, dorsal view; 103, Palpus of male, ventral view; 104, 105, Spermathecae, ventral view; 106, Palpus of male, retrolateral view. *co*, copulatory opening; *con*, conductor; *ct*, copulatory tube; *e*, embolus; *ra*, retrolateral apophysis; *spt*, spermatheca; *teg*, tegulum.

Comments. Individuals of *C. obesa* closely resemble those of *C. mixta*, but differ by the longer embolus and by the large dorsal excavation and broad tip of the retrolateral apophysis of the palpal tibia in males, and by the folded copulatory tubes in females. In addition the posterior indentation of the female epigynal plate is usually deeper and often has a small median prominence.

Range. Manitoba to Newfoundland, southward to Mississippi and North Carolina.

Biology. Individuals of *C. obesa* are usually found on low-growing shrubs in deciduous forests but have also been collected on the trunks of fruit trees and have been swept from tall grass. One specimen was found inhabiting a silk nest of the fall webworm, and another was taken in a pitfall trap in a hayfield. Mature males have been found from April to August, and mature females from May to September. Egg sacs were found in June and July.

Clubiona mixta Emerton

Figs. 107–113; Map 16

Clubiona mixta Emerton, 1890:180, fig. 2–2*b* (pl. 5); Edwards 1958:400, figs. 119, 120, 141, 165, 223.

Male. Total length approximately 6.30 mm; carapace 2.32–3.54 mm long, 1.65–2.27 mm wide (eight specimens measured). Carapace orange. Chelicerae dark orange, with ridge along lateral margin, with shallow depression along anteromesal surface. Legs yellow orange. Abdomen dull orange red, with indistinct orange scutum covering anterior one-third to one-half of dorsum. Patella of palpus with short rounded ventral apophysis. Tibia of palpus with broad retrolateral apophysis having a straight or slightly convex ventral margin, a narrow dorsal excavation, and a pointed tip (Figs. 108, 113). Tegular apophysis flattened, tonguelike; embolus broad at base, arched around tip of tegulum, extending along groovelike conductor less than one-half length of tegulum, slightly bent near tip (Fig. 110).

Female. Total length approximately 7.70 mm; carapace 3.34 ± 0.37 mm long, 2.34 ± 0.28 mm wide (10 specimens measured). General structure and color essentially as in male but chelicerae lacking lateral ridge and anteromesal depression and abdomen lacking dorsal scutum. Epigynal plate with shallow indentation in posterior margin, without median prominence within posterior indentation; copulatory openings conspicuous, cavitylike, located near posterolateral angles of epigynal plate (Figs. 107, 109). Copulatory tubes not folded; spermathecae each in two parts occupying nearly full width of epigynal plate, with anterior part pale, rounded, large and with posterior part darker, elongate, smaller (Figs. 111, 112).

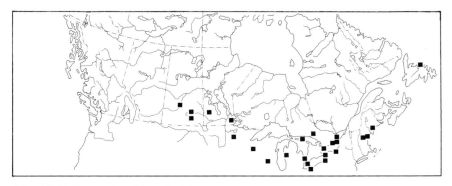

Map 16. Collection localities of *Clubiona mixta*.

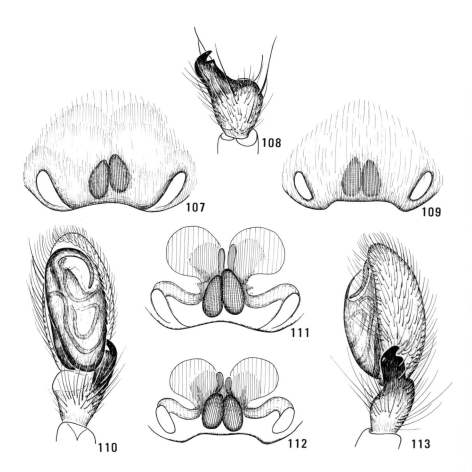

Comments. Individuals of *C. mixta* closely resemble those of *C. obesa* but can be distinguished from the latter by the shorter embolus and by the small excavation and pointed tip on the retrolateral apophysis of the male palpal tibia, and by the lack of a fold in the female copulatory tubes. In addition, the posterior indentation in the epigynal plate is shallower and without a median prominence.

Range. Saskatchewan to Newfoundland, southward to Oklahoma and New Jersey.

Biology. Individuals of *C. mixta* have been collected in the same habitats as *C. obesa* and have also been found occasionally under stones or in plant litter. Mature males have been taken from May to July, and mature females from May to August.

Clubiona spiralis Emerton

Figs. 114−117; Map 17

Clubiona spiralis Emerton, 1909:219, fig. 10 (pl. 10); Edwards 1958:401, figs. 111−113, 143, 175, 212.

Male. Total length approximately 5.80 mm; carapace 2.46−3.19 mm long, 1.62−2.20 mm wide (eight specimens measured). Carapace orange or yellow orange. Chelicerae orange, without longitudinal ridge or depression. Legs yellow orange. Abdomen yellow or orange yellow. Tibia of palpus with strong retrolateral apophysis having convex ventral margin, minute excavation and fingerlike prominence on dorsal margin (Fig. 117). Tegular apophysis long, tapered, extending distad nearly to level of cymbial tip (Fig. 114); embolus slender, with base concealed by tegular apophysis in ventral view, angled near tip, extending basad approximately one-quarter length tegulum (Fig. 114).

Female. Total length approximately 6.10 mm; carapace 2.70−2.95 mm long, 1.88−1.98 mm wide (three specimens measured). General structure and color essentially as in male. Epigynum with large smooth median swelling and with anterolateral crescent-shaped depression bordered anteriorly by recurved dark ridge; copulatory openings slitlike, transverse, located on median swelling of epigynal plate (Fig. 115). Copulatory tubes slender, curving anterolaterad; spermathecae small, in two parts, with anterior part bearing small fingerlike anteromesal prominence (Fig. 116).

Figs. 107−113. Genitalia of *Clubiona mixta*. 107, 109, Epigynums; 108, Tibia of male palpus, dorsal view; 110, Palpus of male, ventral view; 111, 112, Spermathecae, ventral view; 113, Palpus of male, retrolateral view.

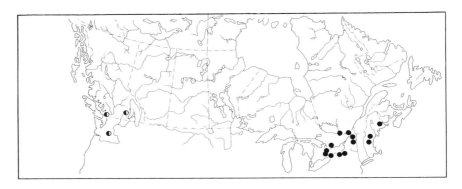

Map 17. Collection localities of *Clubiona spiralis* (●) and *C. mimula* (◖).

Comments. The male of *C. spiralis* somewhat resembles those of *C. obesa* and *C. mixta* but differs in having a much more convex ventral margin on the retrolateral apophysis; the fingerlike prominence on the dorsal margin of the retrolateral apophysis and the more slender embolus are also diagnostic for male *spiralis*. Females of *spiralis* are unique within the group in having slitlike transverse copulatory openings.

Range. Southern Ontario to Maine, southward to New Jersey.

Biology. Both sexes of *C. spiralis* have been collected in pitfall traps or by sweep nets in mixed conifer−deciduous forests. One female was taken on a camp building in such a forest. Adults have been taken in May and August.

Clubiona mimula Chamberlin

Figs. 118−121; Map 17

Clubiona mimula Chamberlin, in Chamberlin & Gertsch, 1928:184; Edwards 1958:397, figs. 103, 104, 129, 138, 178, 229.

Male. Total length approximately 5.40 mm; carapace 2.31−2.58 mm long, 1.60−1.82 mm wide (four specimens measured). Carapace dull orange with darker lines and paler bands radiating from dorsal groove area. Chelicerae dark orange, without ridges or depressions. Legs dull orange or yellow orange. Abdomen dull yellow to dull red. Patella of palpus with short wide apophysis at tip on prolateral side. Tibia of palpus approximately as long as wide, with broad retrolateral apophysis bearing fingerlike terminal tooth on ventral side (Fig. 121). Tegulum convex, with small hooked bilobed apophysis near base of embolus, and with short thick flange mesally; embolus slender, tapered, arched around tip of tegulum, extending basad along membranous groove approximately one-fifth length of tegulum, with base hidden (Fig. 118).

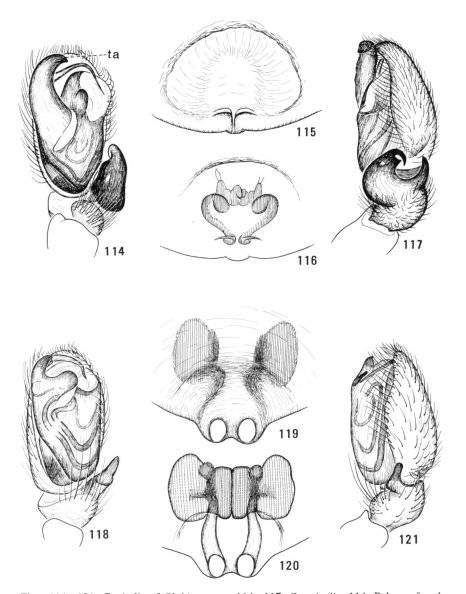

Figs. 114–121. Genitalia of *Clubiona* spp. 114–117, *C. spiralis*. 114, Palpus of male, ventral view; 115, Epigynum; 116, Spermathecae, ventral view; 117, Palpus of male, retrolateral view. 118–121, *C. mimula*. 118, Palpus of male, ventral view; 119, Epigynum; 120, Spermathecae, ventral view; 121, Palpus of male, retrolateral view. *ta*, tegular apophysis.

Female. Total length approximately 5.60 mm; carapace 2.62−2.90 mm long, 1.83−2.00 mm wide (five specimens measured). General structure and color essentially as in male. Epigynum with elongate plate that is extended posteriad over genital groove as paired blunt prominences, with paired deep depressions in posterior half; copulatory openings distinct, cavitylike, slightly separated, located at posterior margin of epigynal plate (Fig. 119). Copulatory tubes rather slender, close together, slightly arched, extending anteriad to level of posterior parts of spermathecae; spermathecae each in two parts that are of different size and arranged transversely (Fig. 120).

Comments. Males of *C. mimula* are distinguished from those of other species in the *obesa* group by the presence of a fingerlike tooth or spur on the retrolateral tibial apophysis. Females are distinguished by the deep depressions in the epigynal plate and by the disparity in size and transverse arrangement of the parts of the spermathecae.

Range. California to British Columbia, eastward to Utah.

Biology. Specimens of *C. mimula* have been collected by sifting leaf litter under deciduous trees and by pitfall traps along the shore of Lake Okanagan. Mature males and females were collected from April to September, and an additional female was taken in February.

Clubiona chippewa Gertsch

Figs. 122−126; Map 18

Clubiona chippewa Gertsch, 1941b:16, figs. 50, 51; Edwards 1958:397, figs. 116−118; Dondale & Redner 1976:1158, figs. 10−15.
Clubiona kuratai Roddy, 1966:405, figs. 11, 12.

Male. Total length approximately 5.10 mm; carapace 2.34, 2.50 mm long, 1.61, 1.80 mm wide (two specimens measured). Carapace orange yellow, slightly suffused with green or gray. Chelicerae yellow orange. Legs pale yellow. Abdomen orange yellow, with indistinct scutum covering most of dorsum. Patella of palpus with short pointed ventral apophysis. Tibia of palpus with strong retrolateral apophysis bearing knoblike process on ventral margin (Figs. 125, 126). Tegular apophysis broad, forming cup around base of embolus (Fig. 124); embolus short, bent at base, extending short distance along tegulum at tip (Fig. 124).

Female. Total length approximately 6.25 mm; carapace 2.80−3.00 mm long, 1.75−2.14 mm wide (three specimens measured). General structure and color essentially as in male but chelicerae relatively shorter and stouter and abdomen lacking dorsal scutum; carapace and chelicerae often dark orange. Epigynum with plate semicircular; copulatory openings located in paired deep lateral pockets (Fig. 122). Copulatory tubes short, extending mesad; spermathecae each in two parts, with the small posterior part round, lying close to genital groove (Fig. 123).

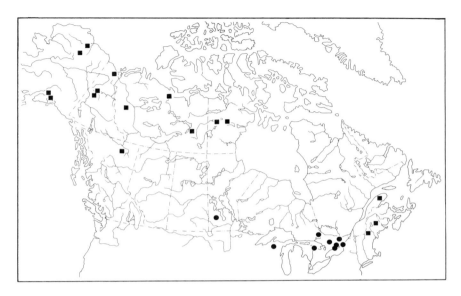

Map 18. Collection localities of *Clubiona chippewa* (●) and *C. praematura* (■).

Comments. The large rounded prominence on the ventral margin of the retrolateral apophysis on the male palpal tibia and the round shape and posterior location of the small parts of the spermathecae distinguish individuals of *C. chippewa* from those of the other species in the *obesa* group.

Range. Manitoba and Wisconsin to southern Ontario.

Biology. Specimens of *C. chippewa* have been collected from meadows and stony ground by pitfall traps. Adult males have been taken in July, and adult females in July and August.

Clubiona furcata Emerton

Figs. 127–131; Map 19

Clubiona furcata Emerton, 1919:106, figs. 8–8c (pl. 7); Edwards 1958:395, figs. 108–110, 161, 162, 220.
Clubiona rowani Gertsch, 1941b:17, fig. 53.

Male. Total length approximately 3.80 mm; carapace 1.75 ± 0.07 mm long, 1.27 ± 0.05 mm wide (13 specimens measured). Carapace yellow orange. Chelicerae dark orange. Legs yellow orange. Abdomen dull red. Patella of palpus with short blunt ventral apophysis. Tibia of palpus with strong retrolateral apophysis arising retrolaterodorsally and curling around to retrolateral side where it terminates in two small teeth (Figs. 130, 131). Tegular apophysis tapered (Fig.

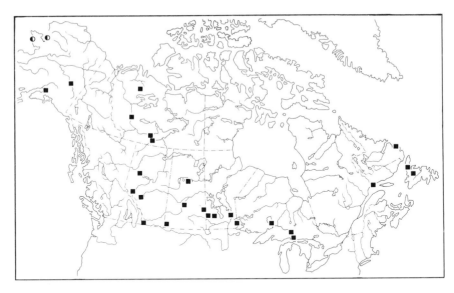

Map 19. Collection localities of *Clubiona furcata* (■) and *C. levii* (◑).

129); embolus arising at distal end of tegular apophysis, with small sclerite at base, with tip extending less than one-quarter length of tegulum (Fig. 129).

Female.   Total length approximately 4.50 mm; carapace 1.87 ± 0.10 mm long, 1.35 ± 0.08 mm wide (14 specimens measured). General structure and color essentially as in male. Epigynum with broad plate indented at midline on posterior margin; copulatory openings small, well-separated, oblique (Fig. 127). Copulatory tubes extending first posterolaterad then anteromesad; spermathecae each in two parts, with posterior part larger, lying directly posteriad of anterior part (Fig. 128).

Comments.   The strong retrolateral apophysis with two teeth, the small sclerite at the base of the embolus, and the small oblique copulatory openings in the epigynum distinguish individuals of *C. furcata* from those of the other members in the *obesa* group.

Range.   Alaska to Newfoundland, southward to Utah.

Biology.   The habitat of *C. furcata* is swampy ground or bogs, and specimens have also been collected from plant litter in tall-grass meadows. One specimen was found in the stomach of a frog (*Rana pipiens*). The egg sac is made within a folded blade of grass. Mature individuals of both sexes have been taken from June to September.

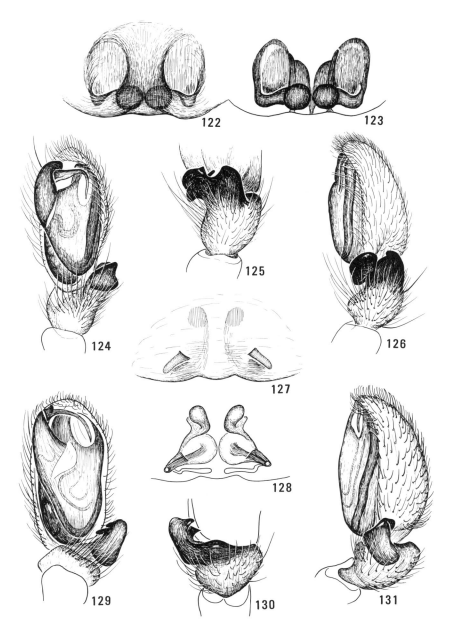

Figs. 122–131. Genitalia of *Clubiona* spp. 122–126, *C. chippewa*. 122, Epigynum; 123, Spermathecae, ventral view; 124, Palpus of male, ventral view; 125, Tibia of male palpus, dorsal view; 126, Palpus of male, retrolateral view. 127–131, *C. furcata*. 127, Epigynum; 128, Spermathecae, ventral view; 129, Palpus of male, ventral view; 130, Tibia of male palpus, dorsal view; 131, Palpus of male, retrolateral view.

Clubiona praematura Emerton

Figs. 132–137; Map 18

Clubiona praematura Emerton, 1909:219, figs. 7–7*b* (pl. 10); Edwards 1958:396, figs. 105–107, 124, 136, 222.

Male. Total length approximately 3.90 mm; carapace 1.80, 1.93 mm long, 1.33, 1.41 mm wide (two specimens measured). Carapace yellow orange, with network of thin brown lines anterior to dorsal groove. Chelicerae orange. Legs yellow orange. Abdomen dull red with several indistinct yellow chevrons on dorsum. Patella of palpus with short blunt ventral apophysis. Tibia of palpus with strong retrolateral apophysis bearing large hollowed prominence on ventral margin (Figs. 136, 137). Tegular apophysis broad, with thin flange on distal half; embolus broad at base, with minute basal tooth, tapering rapidly within arched part of tegulum, extending basad along tegulum approximately one-third length of tegulum (Fig. 134).

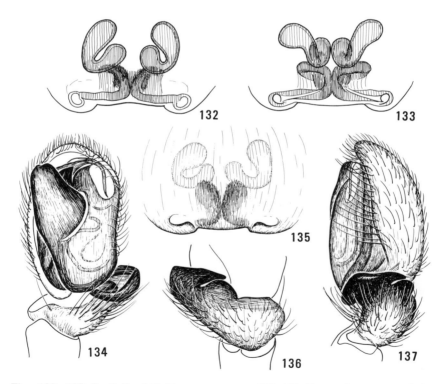

Figs. 132–137. Genitalia of *Clubiona praematura*. 132, 133, Spermathecae, ventral view; 134, Palpus of male, ventral view; 135, Epigynum; 136, Tibia of male palpus, dorsal view; 137, Palpus of male, retrolateral view.

Female. Total length approximately 5.70 mm; carapace 2.03−2.32 mm long, 1.45−1.68 mm wide (seven specimens measured). General structure and color essentially as in male but network of dark lines on carapace often indistinct. Epigynum with broad slightly convex plate having broad indentation in posterior margin; copulatory openings small, located near posterolateral margins of plate (Fig. 135). Copulatory tubes extending mesad along posterior margin of epigynal plate then anteriad together at midline; spermathecae each in two parts that lie in anterior half of epigynum, with parts connected by curved tube, and with posterior part smaller (Figs. 132, 133).

Comments. The large hollowed prominence on the ventral margin of the retrolateral apophysis, the flange on the tegular apophysis, the small female copulatory openings located far laterad, and the copulatory tubes that extend mesad along the posterior margin of the epigynum distinguish individuals of *C. praematura* from those of the other members of the *obesa* group.

Range. Alaska to Maine.

Biology. Specimens of *C. praematura* have been collected in plant litter or under stones. Mature individuals of both sexes have been taken in June and July.

Clubiona bryantae Gertsch

Figs. 138−142, 144; Map 20

Clubiona agrestis Emerton, 1924:144, figs. 6*a*, 6*b*. Name *agrestis* preoccupied in genus *Clubiona*.
Clubiona bryantae Gertsch, 1941*b*:16; Edwards 1958:402, figs. 114, 115, 137, 174, 221.

Male. Total length approximately 4.95 mm; carapace 2.28 ± 0.18 mm long, 1.59 ± 0.14 mm wide (20 specimens measured). Carapace yellow orange to yellow brown. Chelicerae dark orange to orange brown, with ridge along lateral margin. Legs tawny or yellow orange. Abdomen orange yellow or red yellow, with indistinct scutum covering nearly entire dorsum. Femur of palpus with short blunt ventral apophysis. Tibia of palpus with retrolateral apophysis convex along ventral margin, deeply cleft to form two long curved pointed parts (Figs. 142, 144). Tegular apophysis long, with two small teeth on retrolateral margin (Fig. 140); embolus short, little curved, tapered, not lying along tegulum but directed retrolaterodistad (Fig. 140).

Female. Total length approximately 6.25 mm; carapace 2.73 ± 0.28 mm long, 1.89 ± 0.20 mm wide (20 specimens measured). General structure and color essentially as in male but abdomen lacking dorsal scutum. Epigynum with smooth slightly convex plate; plate with broad median prominence marked off by pair of oblique folds, and with small excavation in posterior margin; copulatory openings long, wide, well-separated (Fig. 141). Copulatory tubes broad, extending anteriad, tightly looped upon themselves near point of junction with spermathecae; spermathecae each in two parts, with anterior part rounded and with posterior part elongate and having lateral prominence (Figs. 138, 139).

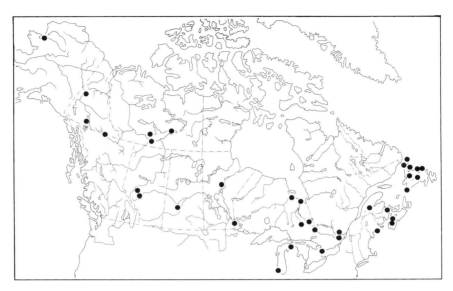

Map 20. Collection localities of *Clubiona bryantae*.

Comments.   Males of *C. bryantae* closely resemble those of *C. levii* but can be distinguished by the long pointed parts of the retrolateral apophysis on the palpal tibia. Females of *bryantae* have shallow well-separated copulatory openings and a narrow median notch in the posterior margin of the epigynal plate.

Range.   Alaska to Newfoundland, southward to Wyoming, Illinois, and Massachusetts.

Biology.   Most of the specimens of *C. bryantae* on hand were collected by pitfall traps placed in hay meadows, or in weedy areas at the margins of deciduous woods. Some were taken in plant litter from spruce−fir forests. Hackman (1954) recorded specimens from herbaceous vegetation in Newfoundland bogs and swamps. Mature individuals of both sexes have been taken from June to August.

Clubiona levii Holm

Figs. 143, 145, 146; Map 19

Clubiona levii Holm, 1960:129, figs. 2−4.

Male.   Total length approximately 5.25 mm; carapace 2.08, 2.43 mm long, 1.50, 1.73 mm wide (two specimens measured). Carapace yellow orange, with network of thin dark lines between dorsal groove and posterior row of eyes.

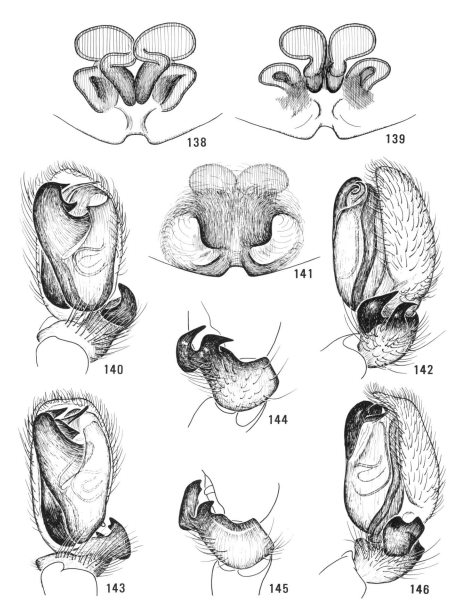

Figs. 138−146. Genitalia of *Clubiona* spp. 138−142, 144, *C. bryantae*. 138, 139, Spermathecae, ventral view; 140, Palpus of male, ventral view; 141, Epigynum; 142, Palpus of male, retrolateral view; 144, Tibia of male palpus, dorsal view. 143, 145, 146, *C. levii*. 143, Palpus of male, ventral view; 145, Tibia of male palpus, dorsal view; 146, Palpus of male, retrolateral view.

Chelicerae brown. Legs yellow. Abdomen red yellow. Femur of palpus with short blunt ventral apophysis. Tibia of palpus with strong retrolateral apophysis having swollen ventral margin and deep narrow cleft forming two blunt points (Figs. 145, 146). Tegular apophysis long, blunt, bearing two teeth on retrolateral margin; embolus short, tapered, little curved, directed retrolaterodistad (Fig. 143).

Female. Unknown.

Comments. Males of *C. levii* closely resemble those of *C. bryantae* but can be distinguished from the latter by the relatively short blunt parts of the retrolateral apophysis on the palpal tibia.

Range. Alaska.

Biology. Holm's type male of *C. levii* was collected in August, and a second male was collected in early September.

The *reclusa* group

Description. Total length 4.90−8.25 mm. Retrolateral apophysis of male palpal tibia strong, bearing two or three teeth; ventral process long, pointed, harpoon-shaped; dorsal process shorter, fingerlike or more simply tapered (e.g., Figs. 147, 149, 161). Tegulum elongate, convex, with tegular apophysis prominent, arising prolaterodistad, bearing one or two sharp teeth at tip; embolus slender and curved, hidden at base (ventral view), with tip lying adjacent to tip of conductor at distal end of alveolus (Figs. 147, 152, 158, 162); conductor prominent, usually tapered (e.g., Figs. 147, 162). Epigynum with broad hairy moderately convex plate having few inconspicuous transverse grooves and having broad prominence at posterior margin (e.g., Figs. 148, 159); copulatory openings slitlike, indistinct, located in posterior margin of epigynal plate. Copulatory tubes usually broad, extending anterolaterad to level of spermathecae where they enlarge and bend abruptly mesad; spermathecae each of two kidney-shaped parts lying close together at midline, with one part dorsal to the other (e.g., Figs. 150, 151, 160).

Comments. The members of the *reclusa* group are distinguished from those of the other groups by the harpoon-shaped ventral part of the retrolateral apophysis on the male palpal tibia, the prominent conductor, and the slitlike copulatory openings located in the posterior margin of the female epigynal plate. Four species are known in Canada.

Key to species of the *reclusa* group

Conductor slender, constricted in ventral view (Fig. 152) ....
.... *pacifica* **Banks** (p. 83)

4(2). Conductor with sides parallel (Fig. 158) .... *kulczynskii* **Lessert** (p. 85)
Conductor slender at base, wider at middle, tapered distad (Fig. 162).........
.... *norvegica* **Strand** (p. 87)

5(1). Epignyal plate with broad three-lobed median prominence at posterior margin
(Figs. 148, 153) ....
.... *canadensis* **Emerton** (p. 81) or *pacifica* **Banks** (p. 83)
Epignyal plate with one- or two-lobed prominence at posterior margin (Figs. 159,
163) 6

6(5). Epignyal plate with one large median prominence at posterior margin (Fig. 159)
.... *kulczynskii* **Lessert** (p. 85)
Epignyal plate with two small median prominences at posterior margin (Fig. 163)
.... *norvegica* **Strand** (p. 87)

Clé des espèces du groupe *reclusa*

1. Mâle 2
Femelle 5

2(1). Conducteur courbé à son extrémité (fig. 147 et 152) 3
Conducteur droit (fig. 158 et 162) 4

3(2). Conducteur large, non rétréci en vue ventrale (fig. 147).................
.... *canadensis* **Emerton** (p. 81)
Conducteur étroit, rétréci en vue ventrale (fig. 152) .. *pacifica* **Banks** (p. 83)

4(2). Conducteur à parois parallèles (fig. 158) .... *kulczynskii* **Lessert** (p. 85)
Conducteur étroit à la base, plus large au milieu et fusiforme à son extrémité
(fig. 162) .... *norvegica* **Strand** (p. 87)

5(1). Plaque épigynale avec large protubérance médiane trilobée à la marge postérieure
(fig. 148 et 153) ....
.... *canadensis* **Emerton** (p. 83) ou *pacifica* **Banks** (p. 81)
Plaque épigynale avec une protubérance unie ou bilobée à la marge postérieure
(fig. 159 et 163) 6

6(5). Plaque épigynale avec une grande protubérance médiane à la marge postérieure
(fig. 159) .... *kulczynskii* **Lessert** (p. 85)
Plaque épigynale avec deux petites protubérances médianes à la marge
postérieure (fig. 163) .... *norvegica* **Strand** (p. 87)

Clubiona canadensis Emerton

Figs. 6, 147−151; Map 21

Clubiona canadensis Emerton, 1890:181, figs. 4−4c (pl. 5); Edwards
1958:406, figs. 150, 151, 155, 163, 216 (in part); Roddy 1973:143, figs. 1−4 (in
part).

Male. Total length approximately 7.05 mm; carapace 3.16 ± 0.23 mm
long, 2.23 ± 0.16 mm wide (20 specimens measured). Carapace yellow orange,
somewhat paler anterior to dorsal groove and with few darker lines radiating from
dorsal groove area. Chelicerae orange brown, without longitudinal ridges. Legs
orange yellow. Abdomen dull red, with inconspicuous scutum covering anterior

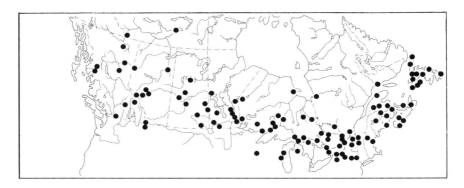

Map 21. Collection localities of *Clubiona canadensis*.

half of dorsum. Tibia of palpus with strong retrolateral apophysis in two processes; ventral process long, harpoon-shaped; dorsal process shorter, fingerlike, with two teeth joined at their bases by low ridge (Fig. 149). Tegular apophysis tapered, twisted at tip, armed with slender tooth (Fig. 147); embolus slender, curved, partly hidden in ventral view by tegular apophysis and conductor; conductor long, tapered, with angular tip (Fig. 147).

Female. Total length approximately 8.25 mm; carapace 3.39 ± 0.37 mm long, 2.41 ± 0.25 mm wide (20 specimens measured). General structure and color essentially as in male but abdomen lacking dorsal scutum and carapace sometimes darkened in eye area (Fig. 6). Epigynum with somewhat convex hairy plate; plate traversed by few grooves, with broad three-lobed median prominence at posterior margin; copulatory openings indistinct, slitlike, located near midline in posterior margin of epigynal plate (Fig. 148). Copulatory tubes extending anterolaterad, then expanding and bending abruptly toward midline; spermathecae each of two kidney-shaped parts, with one part lying immediately dorsal to the other (Figs. 150, 151).

Comments. Specimens of *C. canadensis* closely resemble those of *C. pacifica* but can be distinguished by the broad conductor on the male palpus. The two species can also be distinguished by their partial geographic separation (see Maps 21, 22). Further work is needed to distinguish females of the two species anatomically.

Range. British Columbia and western Northwest Territories to Newfoundland, southward to Colorado and to North Carolina.

Biology. Specimens of *C. canadensis* have been collected from various kinds of trees and shrubs, under loose bark, under stones, and in leaf litter and moss, rarely in houses. Mature males and females have been taken from April to October. Females with egg sacs have been found from July to September.

Clubiona pacifica Banks

Figs. 152–157; Map 22

Clubiona pacifica Banks, 1896:65; Roddy 1973:145, figs. 5–8 (in part).

Male. Total length approximately 6.00 mm; carapace 2.71 ± 0.19 mm long, 1.93 ± 0.13 mm wide (15 specimens measured). Carapace yellow orange, somewhat paler anterior to dorsal groove, with few darker lines radiating from dorsal groove area. Chelicerae orange brown, without longitudinal ridges. Legs orange yellow. Abdomen dull red, with inconspicuous scutum covering anterior half of dorsum. Tibia of palpus with strong retrolateral apophysis in two processes; ventral process long, harpoon-shaped; dorsal process shorter, fingerlike, rather broad at base, joined to base of ventral process by prominent ridge (Fig. 154). Tegular apophysis twisted at tip, armed with slender tooth (Fig. 152); embolus slender, curved, with tip lying near tip of conductor; conductor short, slender, little tapered, with angular tip (Fig. 152).

Female. Total length approximately 6.75 mm; carapace 2.85 ± 0.24 mm long, 2.02 ± 0.18 mm wide (14 specimens measured). General structure and color essentially as in male but abdomen lacking dorsal scutum, and carapace sometimes much darkened in eye area. Epigynum with somewhat convex hairy plate

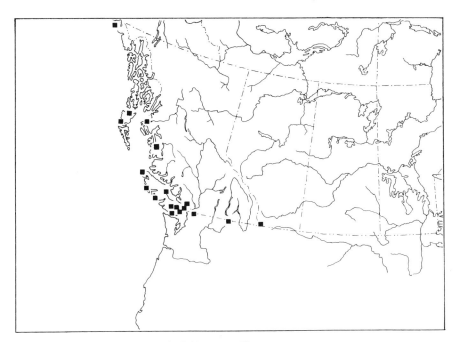

Map 22. Collection localities of *Clubiona pacifica*.

83

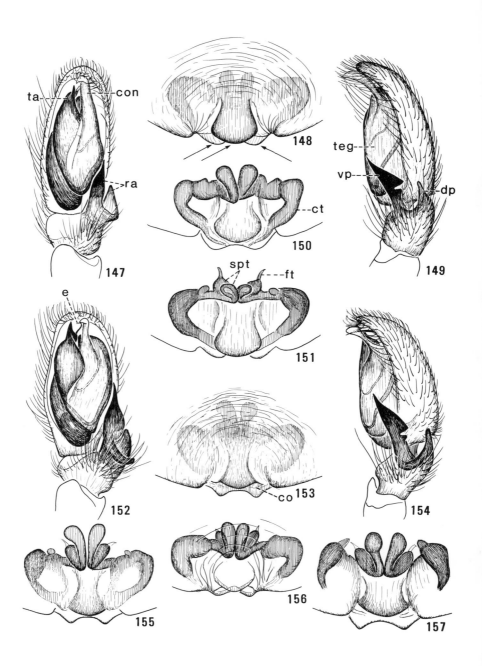

traversed by few grooves and with broad three-lobed prominence at posterior margin; copulatory openings slitlike, inconspicuous, located near midline at posterior margin of plate (Fig. 153). Copulatory tubes extending anterolaterad, then expanded and angled abruptly toward midline; spermathecae each in two kidney-shaped parts, with one part lying immediately dorsal to the other (Figs. 155–157).

Comments. Individuals of *C. pacifica* closely resemble those of *C. canadensis* but are distinguished from the latter by the slender palpal conductor of the male. The two species can also be distinguished by their partial geographic separation (see Maps 21, 22). Further work is needed to distinguish females of these species anatomically.

Range. Coastal southern Alaska to California, inland to southern Alberta.

Biology. Specimens of *C. pacifica* have been collected by sweep nets from grasses or shrubs in fir forests, subalpine meadows, salt marshes, and lake shores. One specimen was found in a lakeside cottage. Mature males have been taken from May to November, and mature females from May to September.

Clubiona kulczynskii Lessert

Figs. 158–161; Map 23

Clubiona kulczynskii Lessert, 1905:647, fig. 13; Edwards 1958:403, figs. 148, 149, 158, 168, 217.
Clubiona intermontana Gertsch, 1933:9, figs. 10, 13.
Clubiona altana Gertsch, 1941b:16, fig. 54.

Male. Total length approximately 4.90 mm; carapace 2.23 ± 0.15 mm long, 1.56 ± 0.09 mm wide (20 specimens measured). Carapace yellow orange, paler anterior to dorsal groove, with few indistinct darker lines radiating from dorsal groove area. Chelicerae dark orange, with longitudinal ridge along anteromesal surface. Legs orange yellow. Abdomen orange or dull red, darker at heart mark, with inconspicuous scutum covering anterior half of dorsum. Patella of palpus concave on ventral surface and with small apophysis at prolaterodistal angle. Tibia of palpus with hairy prominence on ventral side, with strong three-part retrolateral apophysis (Fig. 161); ventral process of retrolateral apophysis harpoon-shaped; middle process short, knobbed; dorsal process broad,

---

Figs. 147–157. Genitalia of *Clubiona* spp. 147–151, *C. canadensis*. 147, Palpus of male, ventral view; 148, Epigynum; 149, Palpus of male, retrolateral view; 150, 151, Spermathecae, ventral view. 152–157, *C. pacifica*. 152, Palpus of male, ventral view; 153, Epigynum; 154, Palpus of male, retrolateral view; 155–157, Spermathecae, ventral view. *co*, copulatory opening; *con*, conductor; *ct*, copulatory tube; *dp*, dorsal process; *e*, embolus; *ft*, fertilization tube; *ra*, retrolateral apophysis; *spt*, spermatheca; *ta*, tegular apophysis; *teg*, tegulum; *vp*, ventral process.

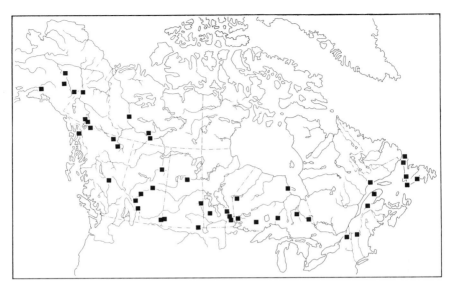

Map 23. Collection localities of *Clubiona kulczynskii*.

hollowed. Tegular apophysis bearing two teeth near base of embolus (Fig. 158); embolus slender, tapered, curved, with tip lying against elongate slightly sclerotized conductor at distal end of alveolus (Fig. 158); conductor uniform in width, flattened at tip, not bent (Fig. 158).

Female.   Total length approximately 6.20 mm; carapace 2.58 ± 0.18 mm long, 1.83 ± 0.13 mm wide (20 specimens measured). General structure and color essentially as in male but chelicerae relatively shorter and stouter. Abdomen without dorsal scutum. Epigynum with broad semicircular slightly convex hairy plate marked by inconspicuous transverse grooves and with median prominence at posterior margin; copulatory openings small, inconspicuous, located in lateral margins of median prominence (Fig. 159). Copulatory tubes slender, extending anteriad then far laterad where they enlarge and bend toward midline; spermathecae each in two parts, with one part lying immediately dorsal to the other (Fig. 160); ventral parts of the two spermathecae partly visible through integument of epigynal plate.

Comments.   The uniform width of the palpal conductor in males and the single large median prominence on the posterior margin of the epigynal plate distinguish specimens of *C. kulczynskii* from those of the other members of the *reclusa* group. Doubt exists regarding the identity of North American specimens with European ones; the former match Tullgren's (1946) illustrations of Swedish material but differ in the external genitalia of both sexes as illustrated by Wiehle (1965). Further work is needed.

Range.   Alaska to Newfoundland, southward to Colorado and to North Carolina; Europe and Asia.

Biology. Specimens of *C. kulczynskii* have been collected from the foliage of lodgepole pines, junipers, spruces, and various deciduous shrubs, in leaf litter in coniferous and aspen forests, and in sphagnum bogs. Mature males have been taken from June to August, and mature females from June to September. Females were found with egg sacs in rolled leaves of elder, willow, and other deciduous trees or shrubs.

Clubiona norvegica Strand

Figs. 162–165; Map 24

Clubiona norvegica Strand, 1900:30, fig. *e*; Edwards 1958:405, figs. 146, 147, 156, 164, 215.
Clubiona carpenterae Fox, 1938:240, fig. 7 (pl. 2).

Male. Total length approximately 4.80 mm; carapace 2.24 ± 0.18 mm long, 1.58 ± 0.17 mm wide (20 specimens measured). Carapace yellow orange, paler anterior to dorsal groove, with few dark lines radiating from dorsal groove area. Chelicerae dark orange or orange brown, each with two longitudinal ridges along anterior surface. Legs orange yellow to orange. Abdomen pale orange to dull red, sometimes darker posteriad, lacking dorsal sclerite. Tibia of palpus with prominent three-part retrolateral apophysis; ventral process long, harpoon-shaped; retrolateral process short, blunt; dorsal process short, pointed (Fig. 165). Tegular apophysis broad, fluted, truncate at tip, provided with sharp tooth at ventral angle

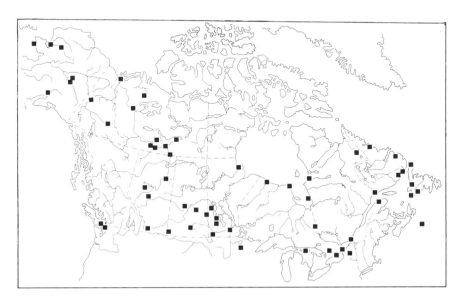

Map 24. Collection localities of *Clubiona norvegica*.

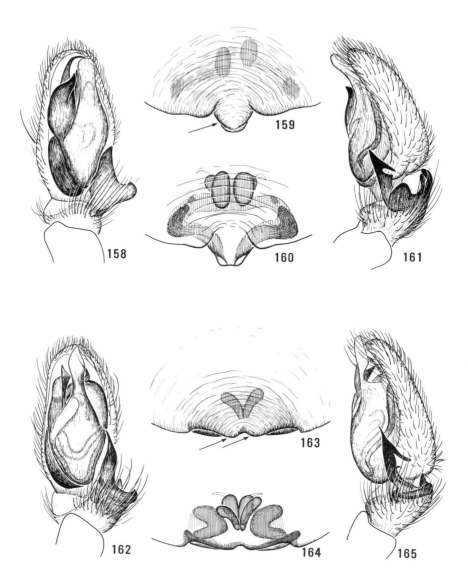

Figs. 158–165. Genitalia of *Clubiona* spp. 158–161, *C. kulczynskii*. 158, Palpus of male, ventral view; 159, Epigynum; 160, Spermathecae, ventral view; 161, Palpus of male, retrolateral view. 162–165, *C. norvegica*. 162, Palpus of male, ventral view; 163, Epigynum; 164, Spermathecae, ventral view; 165, Palpus of male, retrolateral view.

(Fig. 162); embolus thin, flat, angled near base, with tip lying along flat slightly sclerotized conductor at distal end of tegulum (Fig. 162). Conductor slender at base, expanded at middle, tapered distad.

Female. Total length approximately 6.65 mm; carapace 2.62 ± 0.23 mm long, 1.88 ± 0.18 mm wide (20 specimens measured). General structure and color essentially as in male but carapace sometimes very dark in eye area, and chelicerae without longitudinal ridges. Epigynum with broad semicircular slightly convex hairy plate traversed by several fine grooves; posterior margin with two small prominences separated by small indentation; copulatory openings inconspicuous, well-separated, located in posterior margin of epigynal plate (Fig. 163). Copulatory tubes broad, extending laterad, then mesad and anterolaterad; spermathecae each in two kidney-shaped parts, one lying immediately dorsal to the other (Fig. 164).

Comments. Specimens of *C. norvegica* are distinguished from those of other members of the *reclusa* group by the conductor of the male palpus, which is slender at the base, expanded at the middle, and tapered distad, and by the presence of two small prominences on the posterior margin of the epigynal plate.

Range. Alaska to Newfoundland, southward to Utah and to New York; Europe.

Biology. Specimens of *C. norvegica* have been collected in sphagnum bogs, beach grasses, and salt marshes, on buildings, rocky lake shores, at the margins of prairie sloughs, and occasionally from the foliage of deciduous shrubs or Douglas firs. Adults of both sexes have been taken from May to September.

The *lutescens* group

Description. Total length 5.70−6.00 mm. Retrolateral apophysis of male palpus with two processes, both prominent, well-separated from base to tip (Figs. 169, 173). Tegulum elongate, convex, with tegular apophysis passing dorsal to embolus (and partly concealed by it in ventral view, as in Figs. 166, 170); embolus prominent, arising at middle or in distal half of tegulum, passing prolaterodistad then arched dorsad and retrolaterad along wall of alveolus, coiled at tip (Figs. 166, 170); conductor absent. Epigynum of female with convex plate; plate without transverse grooves or ridges, projecting posteriad over genital groove in pair of blunt prominences; copulatory openings conspicuous, located in posterior part of plate (Figs. 167, 171). Copulatory tubes short, broad, extending laterad and/or anteriad; spermathecae in one or two parts (Figs. 168, 172).

Comments. The fully exposed broadly arched embolus with a distal coil, and the broad copulatory tubes of females, distinguish specimens of the *lutescens* group from those of the other groups. Two species are known in Canada.

Key to species of the *lutescens* group

1. Male 2
   Female 3
2(1). Retrolateral apophysis of palpal tibia with ventral process longer and more slender than dorsal process (Fig. 169). Embolus tip hardly exposed in retrolateral view (Fig. 169) .... *riparia* **L. Koch** (p. 90)
   Retrolateral apophysis of palpal tibia with ventral process shorter and approximately equal in thickness to dorsal process (Fig. 173). Embolus tip well exposed in retrolateral view (Fig. 173) ..... *lutescens* **Westring** (p. 92)
3(1). Copulatory openings close together, elongate, extending to posterior margin of epigynal plate (Fig. 167) .... *riparia* **L. Koch** (p. 90)
   Copulatory openings well-separated, funnellike, not extending to posterior margin of epigynal plate (Fig. 171) .... *lutescens* **Westring** (p. 92)

Clé des espèces du groupe *lutescens*

1. Mâle 2
   Femelle 3
2(1). Apophyse rétrolatérale du tibia palpal avec processus ventral plus long et plus étroit que le processus dorsal (fig. 169). Extrémité de l'embolus à peine visible en vue rétrolatérale (fig. 169) .... *riparia* **L. Koch** (p. 90)
   Apophyse rétrolatérale du tibia palpal avec processus ventral plus court et d'épaisseur à peu près égale au processus dorsal (fig. 173). Extrémité de l'embolus bien visible en vue rétrolatérale (fig. 173) ....
   .... *lutescens* **Westring** (p. 92)
3(1). Ouvertures copulatoires rapprochées, allongées, se prolongeant à la marge postérieure de la plaque épigynale (fig. 167) .·. *riparia* **L. Koch** (p. 90)
   Ouvertures copulatoires bien séparées, infundibuliformes, ne se prolongeant pas à la marge postérieure de la plaque épigynale (fig. 171) ....
   .... *lutescens* **Westring** (p. 92)

Clubiona riparia L. Koch

Figs. 1–3, 5, 166–169; Map 25

Clubiona riparia L. Koch, 1866:293, fig. 187 (pl. 12); Edwards 1958:430, figs. 125–128, 145, 219.
Clubiona ornata Emerton, 1890:183, figs. 9, 9a (pl. 5). Name *ornata* preoccupied in genus *Clubiona*.
Clubiona americana Banks, 1892:22.

Male. Total length approximately 5.45 mm; carapace 2.55 ± 0.25 mm long, 1.79 ± 0.17 mm wide (20 specimens measured). Carapace yellow orange to dark orange, often with several darker streaks radiating from dorsal groove area. Chelicerae dark orange, each with two ridges along anterior surface and with depression along mesal surface. Legs orange. Abdomen dull red, with darker broken median band and paler submedian bands, with indistinct scutum covering

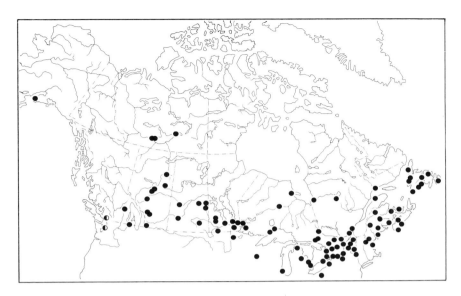

Map 25. Collection localities of *Clubiona riparia* (●) and *C. lutescens* (◐).

approximately one-half of dorsum. Patella of palpus with blunt apophysis at prolaterodistal angle. Tibia of palpus with retrolateral apophysis with two processes; ventral process broad basad, slender distad; dorsal process shorter, less pointed (Fig. 169). Tegular apophysis inconspicuous, ridgelike, arising near base of, and passing dorsal to, embolus (Fig. 166); embolus arising broadly near middle of tegulum, extending prolaterodistad then arched dorsad and retrolaterad along wall of alveolus (Fig. 166), with tip hardly exposed in retrolateral view (Fig. 169).

Female. Total length approximately 7.20 mm; carapace 2.84 ± 0.22 mm long, 2.05 ± 0.17 mm wide (20 specimens measured). General structure and color essentially as in male but chelicerae lacking ridges and depressions and abdomen lacking dorsal scutum. Epigynum with moderately convex plate; plate lacking transverse grooves and ridges, hairy, projecting posteriad over genital groove as two blunt prominences; copulatory openings conspicuous, somewhat angular, close together at midline near posterior margin of plate (Fig. 167). Copulatory tubes broad, short, extending anteriad; spermathecae each in two parts, with one dorsal to the other, with ventral part kidney-shaped and with dorsal part curled laterad at anterior end (Fig. 168).

Comments. Specimens of *C. riparia* can be distinguished from those of *C. lutescens* by the longer, more slender ventral process of the retrolateral apophysis and the less exposed embolus tip in males, and by the closely spaced copulatory openings in females.

Range. Alaska to Newfoundland, southward to New Mexico and to Maryland.

91

Biology.   Specimens of *C. riparia* have been collected from tall grass in marshes and near sloughs and lakes, both by sweep nets and pitfall traps. A few were collected by pitfall traps in mixed conifer–deciduous forests. The egg sacs are made in folded grass blades, and have been found in June and August. Adult males have been taken from May to July, and adult females from May to September.

Clubiona lutescens Westring

Figs. 170–173; Map 25

Clubiona lutescens Westring, 1851:49; Roddy 1966:405, figs. 13–16.
Clubiona assimilata O. Pickard-Cambridge, 1862:7947.

Male.   Total length approximately 6.10 mm; carapace 2.85 mm long, 2.18 mm wide (one specimen measured). Carapace yellow orange, with a few darker bands radiating from dorsal groove. Chelicerae orange brown, with ridge along anterior surface and with shallow groove along anteromesal surface. Legs yellow orange. Abdomen dull red, with inconspicuous darker heart mark, with inconspicuous scutum covering anterior one-third of dorsum. Patella of palpus with flat apophysis at prolaterodistal extremity. Tibia of palpus with retrolateral apophysis with two processes; ventral process short, flat; dorsal process much longer, darker, somewhat hollowed on mesal side (Fig. 173). Tegular apophysis slender, not prominent, lying dorsal to embolus; embolus arising at retrolateral margin of tegulum, extending prolaterodistad, arched dorsad and retrolaterad along wall of alveolus (Fig. 170), with tip well exposed in retrolateral view (Fig. 173).

Female.   Total length approximately 5.70 mm; carapace 2.15 mm long, 1.43 mm wide (one specimen measured). General structure and color essentially as in male but carapace often darker in eye region, chelicerae lacking ridge and depression, and abdomen lacking dorsal scutum. Epigynum with large plate; plate somewhat convex, without transverse ridges or grooves, with broad bilobed prominence at posterior margin; copulatory openings funnellike, well-separated, located posterolaterally at ends of low shiny transverse ridge (Fig. 171). Copulatory tubes broad, extending first laterad then anteriad, bent abruptly mesad; each spermatheca of one part, approximately kidney-shaped, with fingerlike process at posterior end (Fig. 172).

Figs. 166–173. Genitalia of *Clubiona* spp. 166–169, *C. riparia*. 166, Palpus of male, ventral view; 167, Epigynum; 168, Spermathecae, ventral view; 169, Palpus of male, retrolateral view. 170–173, *C. lutescens*. 170, Palpus of male, ventral view; 171, Epigynum; 172, Spermathecae, ventral view; 173, Palpus of male, retrolateral view. *co*, copulatory opening; *ct*, copulatory tube; *cym*, cymbium; *dp*, dorsal process; *e*, embolus; *pat*, patella; *ra*, retrolateral apophysis; *spt*, spermatheca; *ta*, tegular apophysis; *vp*, ventral process.

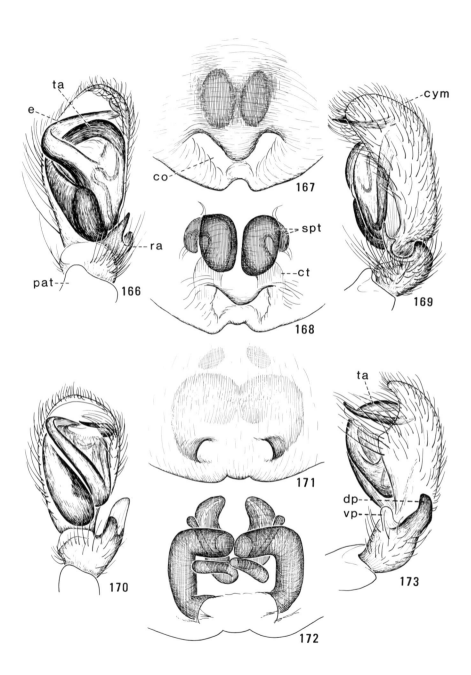

Comments.   Males of *C. lutescens* are distinguished from those of *C. riparia* by having the ventral process of the retrolateral tibial apophysis much shorter than the dorsal process and having a well-exposed embolus tip. Females of *lutescens* differ from those of *riparia* by having the copulatory tubes well separated and funnellike rather than close together and elongate. Specimens were first found in North America in 1949, and the species is believed to be introduced from Eurasia.

Range.   British Columbia and Washington; Europe and Asia.

Biology.   In Britain, *C. lutescens* is found in "moist places, grasses and bushes and trees in woods," and is adult in May and June (Locket and Millidge 1951). Roddy (1966) reports adults in the Vancouver and Seattle areas in June and August. Egg sacs have been found in rolled leaves on willows and similar trees in British Columbia.

The *pallidula* group

Description.   Total length 7.00−10.10 mm. Retrolateral apophysis of male palpus stout, with cluster of processes approximately equal in length, and with roomy excavation within cluster (Figs. 174, 177). Tegulum somewhat convex, with stout apophysis that bears one tooth (Fig. 174); embolus short, stout, curved, arising at tip of tegular apophysis, extending retrolaterad approximately one-half width of tegulum, free at tip (Fig. 174). Epigynum of female with convex rugose plate, excavated at lateral and posterior margins; copulatory openings large, shallow, ovoid, situated near lateral margins of epigynal plate, connected to posterior margin of plate by broad shallow grooves (Fig. 175). Copulatory tubes slender, arched laterad, arising near posterior margin of epigynum; spermathecae each in two parts, with dark ovoid part lying near midline and paler elongate part lying laterally (Fig. 176).

Comments.   The cluster of processes of nearly equal length on the retrolateral apophysis and the short curved embolus of males, and the rugose epigynal plate, large shallow copulatory openings, and large spermathecae, the mesal parts of which are ovoid in outline, in females, distinguish specimens of the *pallidula* group from those of the other groups in *Clubiona*. One species is known in Canada.

Clubiona pallidula (Clerck)

Figs. 174–177; Map 26

Araneus pallidulus Clerck, 1757:81, fig. 7 (pl. 2).
Aranea holosericea Linnaeus, 1758:622.
Aranea amarantha Walckenaer, 1802:219.
Aranea aloma Walckenaer, 1802:219.
Aranea epimelas Walckenaer, 1802:219.
Clubiona incomta C.L. Koch, 1837:19, fig. 442.
Clubiona pallidula: Thorell, 1856:97; Locket & Millidge 1951:133, figs. 66*a*, 66*b*; Roddy 1966:406, figs. 17–20.
Clubiona formosa Blackwall, 1861:125, fig. 78 (pl. 7).

Male. Total length approximately 7.25 mm; carapace 3.15–3.59 mm long, 2.21–2.53 mm wide (three specimens measured). Carapace yellow orange or yellow brown. Chelicerae orange brown, each with depression along anteromesal surface. Legs yellow. Abdomen dull red. Patella of palpus with small apophysis at prolaterodistal angle. Tibia of palpus with stout retrolateral apophysis bearing small sharp process near base on dorsal side, broad process close to tip, and slender fingerlike ventral process (Fig. 177). Tegulum somewhat convex, with dark pointed tooth at tip near base of embolus (Fig. 174); embolus short, curved, arising at tip of tegular apophysis, extending retrolaterad approximately one-half width of tegulum, free at tip (Fig. 174).

Female. Total length approximately 10.10 mm; carapace 3.57, 3.86 mm long, 2.47, 2.65 mm wide (two specimens measured). General structure and color essentially as in male but chelicerae lacking depressions along anteromesal surfaces. Epigynum with convex rugose plate; copulatory openings large, shallow, ovoid, situated near lateral margins of epigynal plate, connected to posterior margin of epigynal plate by broad shallow grooves (Fig. 175).

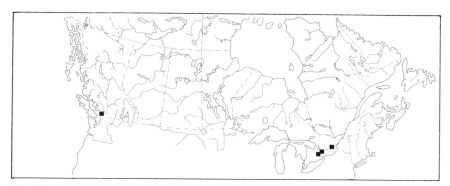

Map 26. Collection localities of *Clubiona pallidula*.

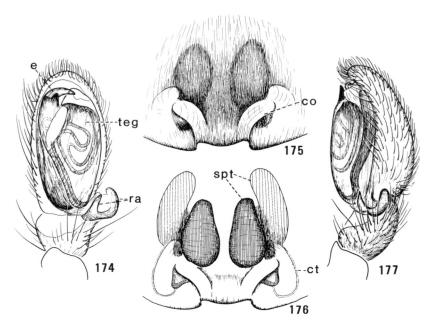

Figs. 174–177. Genitalia of *Clubiona pallidula*. 174, Palpus of male, ventral view; 175, Epigynum; 176, Spermathecae, ventral view; 177, Palpus of male, retrolateral view. *co*, copulatory opening; *ct*, copulatory tube; *e*, embolus; *ra*, retrolateral apophysis; *spt*, spermatheca; *teg*, tegulum.

Copulatory tubes slender, arched laterad, arising near posterior margin of epigynum; spermathecae each in two parts, with dark ovoid mesal part and paler elongate lateral part (Fig. 176).

Comments. Males of *C. pallidula* differ from those of all other species of North American *Clubiona* by having a cluster of processes of nearly equal length on the retrolateral apophysis and a short curved free embolus. Females of *pallidula* differ by having a rugose epigynal plate with large shallow copulatory openings, and by having the mesal parts of the spermathecae large and ovoid.

Range. Coastal British Columbia and Washington, and southern Ontario; Europe, Asia.

Biology. *C. pallidula* is believed to be a recent introduction to North America, as the earliest record of it in this continent is dated 1949 (Roddy 1966). Specimens have been collected on shrubs, herbs, under bark, in mole burrows in the soil, in cultivated crops such as grape vines, and in houses. Adult males have been found from April to June, and females from June to November. Gerhardt (1924) described the mating behavior.

Genus *Clubionoides* Edwards

Spiders of the genus *Clubionoides* spend the daylight hours in silken sacs, then emerge and forage at night. Their claw tufts and leg scopulae provide a secure footing on plant surfaces. Their bodies are various hues of yellow, pale orange, or gray.

Description. Total length 5.85−6.65 mm. Carapace (Fig. 7) yellow orange, ovoid in dorsal view, highest at level of dorsal groove, with sparse covering of pale recumbent setae, and with dark shallow dorsal groove. Eyes nearly uniform in size, arranged in two transverse rows; anterior row recurved; posterior row straight or slightly procurved, slightly longer than anterior row; posterior median eyes slightly closer to posterior lateral eyes than to each other. Chelicerae yellow, long, slender, with four or five teeth on promargin of fang furrow and with four teeth on retromargin. Palp-coxal lobes distinctly longer than wide, constricted at middle on lateral margin. Legs yellow, long, slender, with dense claw tufts and rather sparse scopulae; femur I with two or three prolateral macrosetae; leg II of male without modified macrosetae; trochanter IV with shallow ventral notch at distal end. Abdomen elongate oval, gray to yellow, with indistinct dorsal scutum in male, with pattern of dark spots and chevrons, with covering of short pale semi-erect setae, and with cluster of long erect curved setae at anterior end (Fig. 7). Tibia of male palpus (Fig. 181) approximately twice as long as wide, with thin flat rounded retrolateral apophysis; cymbium lacking spur; tegulum without apophysis, sclerotized only on prolateral half; embolus short, blunt, arising at distal end of tegulum. Epigynum of female (Fig. 179) with concave plate and distinct median scape; copulatory openings small, located at posterior end or at lateral margins of median septum. Copulatory tubes short, slender, straight or curved; spermathecae simple and saclike or in two parts, rather small, situated close together near posterior end of epigynum (Fig. 180).

Comments. Specimens of *Clubionoides* spp. closely resemble those of *Clubiona* spp. but differ from the latter in having two or three prolateral macrosetae (rather than one) on femur I. The minute blunt terminal embolus and partly sclerotized tegulum of males, and the presence of a distinct scape in the epigynum of females, are also diagnostic. Specimens of *Clubionoides* spp. differ from those of *Cheiracanthium* spp. in having a conspicuous dorsal groove, more than two teeth on the promargin of the cheliceral fang furrow, a cluster of long erect setae at the anterior end of the abdomen, and no ventral macroseta at the distal end of basitarsus I. The presence of a lateral constriction on the palp-coxal lobes distinguishes specimens of *Clubionoides* spp. from those of the remaining genera of Clubionidae represented in Canada.

The genus *Clubionoides* accommodates five species in North America north of Mexico and a larger but unknown number in Central and South America (Edwards 1958). A single species is found in Canada.

Clubionoides excepta (L. Koch)

Figs. 7, 178–181; Map 27

Clubiona pallens Hentz, 1847:449, fig. 13 (pl. 23). Name *pallens* preoccupied in genus *Clubiona*.

Clubiona excepta L. Koch, 1866:300, fig. 191 (pl. 22).

Clubionoides excepta: Edwards, 1958:377, figs. 19, 31–33, 211.

Male. Total length approximately 5.85 mm; carapace 2.80 ± 0.20 mm long, 1.98 ± 0.16 mm wide (19 specimens measured). Carapace yellow orange. Chelicerae yellow, without ridges or depressions. Legs yellow. Abdomen pale gray to pale yellow, with approximately five pairs of dark lateral spots, with approximately eight slender chevrons, and with indistinct scutum covering anterior part of dorsum; venter with pair of dark spots anterior to spinnerets. Tibia of palpus with thin flat rounded retrolateral apophysis bearing small point on dorsal margin (Fig. 181). Tegulum sclerotized only on prolateral half, with small hooklike apophysis on retrolateral half (Fig. 178); embolus small, angular, arising at distal end of tegulum (Fig. 178).

Female. Total length approximately 6.65 mm; carapace 2.67–3.35 mm long, 1.85–2.36 mm wide (nine specimens measured). General structure and color essentially as in male but abdomen lacking dorsal scutum. Epigynum with elongate plate; plate depressed mesally, with numerous transverse grooves and ridges, and with prominent scape; copulatory openings located lateral to scape (Fig. 179). Copulatory tubes short; spermathecae each of two parts, with slender bulbous anterior part and round posterior part located near posterior margin of epigynal plate (Fig. 180).

Comments. The sclerotization of only the prolateral half of the tegulum in the male palpus and the presence of a distinct scape in the epigynum of the female distinguish specimens of *C. excepta* from those of the other clubionid genera in Canada.

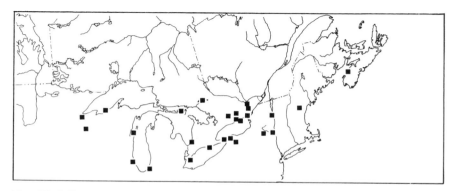

Map 27. Collection localities of *Clubionoides excepta*.

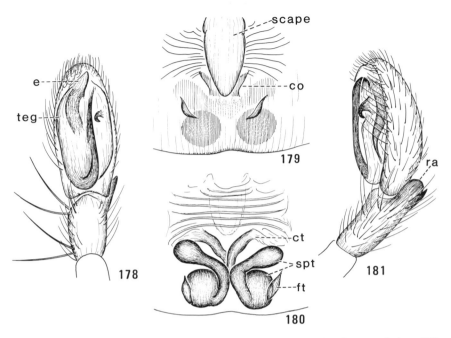

Figs. 178–181. Genitalia of *Clubionoides excepta*. 178, Palpus of male, ventral view; 179, Epigynum; 180, Spermathecae, ventral view; 181, Palpus of male, retrolateral view. *co*, copulatory opening; *ct*, copulatory tube; *e*, embolus; *ft*, fertilization tube; *ra*, retrolateral apophysis; *spt*, spermatheca; *teg*, tegulum.

Range. Nebraska and Minnesota to Nova Scotia, southward to Texas and the West Indies.

Biology. Specimens of *C. excepta* have been collected under loose bark of trees, among dead leaves on the ground in deciduous forests, and under stones. Mature males have been found from late April to August, and mature females from May to October. Egg sacs are spherical, constructed of loose silk threads, and usually covered with soil particles.

Genus *Castianeira* Keyserling

Spiders of the genus *Castianeira* have elongate antlike bodies and rather long thin legs. The carapace and abdomen are heavily sclerotized and shiny. The colors are orange or yellow with contrasting brown and black; the carapace and leg femora may have a bluish iridescence, and the dorsum of the abdomen often shows transverse bands of white scalelike setae. The movements of these spiders are decidedly antlike, and involve rapid darts into the open and equally rapid retreats into crevices or other cover, frequent jerky pauses and changes of direction, bobbing of the abdomen, and the holding of the front legs as if they were an ant's antennae. These spiders are usually found in leaf litter in shady deciduous forests or under logs and stones in more open habitats. They are commonly found in association with ants, and presumably derive some survival advantage through mimicry. The egg sacs are shiny disks with adhering crumbs of soil, stuck to the undersides of logs or stones.

Description. Total length 4.35−8.00 mm. Carapace (Fig. 10) elongate ovoid, bulging, highest at or anterior to dorsal groove, orange to dark orange brown or black, usually darker in eye area, with distinct dorsal groove, with short sparse scalelike recumbent setae, and with few longer erect slender setae on front and eye area. Eyes (Fig. 10) small, uniform in size, in two transverse rows; anterior row slightly recurved, with medians slightly closer to laterals than to each other; posterior row distinctly procurved, slightly longer than anterior row, with medians closer to laterals than to each other. Chelicerae moderately long, stout, hairy, orange to dark brown or black, with two teeth on promargin of fang furrow, two (rarely three) teeth on retromargin. Palp-coxal lobes convex, not constricted on lateral margins. Legs rather long, slender, orange to dark brown or black, with short dense dark claw tufts and thin scopulae; trochanter IV with distinct notch at tip on ventral side (Fig. 9); femur I with two or three dorsal macrosetae, one or two prolaterals; basitarsus I with two pairs of ventral macrosetae, without median ventral macroseta at tip. Abdomen (Fig. 10) elongate ovoid, rather slender, often with transverse bands of white scalelike setae, with large shiny dorsal, epigastric, and ventral scuta (reduced in females), without cluster of long erect setae at anterior end. Femur and patella of male palpus without apophyses. Tibia of male palpus as long as wide or only slightly longer than wide, concave prolateromesally, with strong retrolateroventral ridge bearing one or more small teeth (e.g., Figs. 182, 192, 201). Cymbium rounded at base, long and slender distally (e.g., Fig. 187). Tegulum rounded at base, long and slender distally, enclosing two loops of seminal duct within base; embolus usually long, slender, straight, spirally twisted at tip, arising directly on elongate tip of tegulum (e.g., Fig. 187). Epigynum of female with rounded convex hairy plate; copulatory openings small and round or elliptical and slitlike, well-separated, distinct, often connected by shallow transverse groove (Fig. 183). Copulatory tubes short to moderately long, slender or moderately broad, usually extending anteriad or mesad; spermathecae longer than wide, touching at midline, rugose, usually with slender posterior part situated short distance anterior to genital groove (e.g., Figs. 185, 194, 205).

Comments. Specimens of *Castianeira* spp. can be distinguished from those of *Cheiracanthium* spp., *Clubiona* spp., and *Clubionoides* spp. by the lack of lateral constrictions at the middle of the palp-coxal lobes, and from those of *Trachelas* spp., *Phrurotimpus* spp., and *Scotinella* spp. by having the posterior row of eyes procurved, by the notched trochanter IV, and by the presence of only two pairs of ventral macrosetae on basitarsus I. Specimens of *Castianeira* spp. differ from those of *Agroeca* spp. by having antlike bodies and legs, by having two pairs of ventral macrosetae on basitarsus I, and by lacking a cluster of long curved erect setae at the anterior end of the abdomen.

Reiskind (1969) revised the North and Central American species of *Castianeira*. He estimated a total of 118 world species, of which 35 occur in North America. Eight species are represented in Canada.

Key to species of *Castianeira*

1. Male 2
   Female 9
2(1). Embolus long, slender, without large teeth (e.g., Figs. 187, 211, 220). Cymbium without dorsal macroseta 3
   Embolus bulbous, with large teeth (Figs. 182, 184). Cymbium with dorsal macroseta near retrolateral margin .... *trilineata* (**Hentz**) (p.104)
3(2). Femora I and II with dark brown longitudinal bands on dorsal, prolateral, and retrolateral surfaces (Fig. 190) .... *cingulata* (**C.L. Koch**) (p.105)
   Femora I and II without dark longitudinal bands 4
4(3). Embolus short (length embolus ÷ length tegulum + embolus × 100 = 13 or less) 5
   Embolus longer (length embolus ÷ length tegulum + embolus × 100 = 15 or more) 6
5(4). Carapace 1.75−2.40 mm wide. Dorsum of abdomen with four or more transverse bands of white setae distributed on both anterior and posterior parts (as in Fig. 10)............................ *longipalpa* (**Hentz**) (p.107)
   Carapace 1.20−1.60 mm wide. Dorsum of abdomen with three or four pale transverse bands on anterior half (as in Fig. 198) .... *gertschi* **Kaston** (p.109)
6(4). Femora I and II (as well as III and IV) dark orange brown with contrasting yellow or orange tips .... *variata* **Gertsch** (p.111)
   Femora I and II without contrasting yellow or orange tips 7
7(6). Embolus long (length embolus ÷ length tegulum + embolus × 100 = 25−26). Dorsum of abdomen without red area posteriorly, with four or five transverse bands of black setae .... *alteranda* **Gertsch** (p.112)
   Embolus less long (length embolus ÷ length tegulum + embolus × 100 = 15−17). Dorsum of abdomen with red area posteriorly (sometimes yellow in alcohol), with transverse bands of white setae (Figs. 212, 214, 221) 8
8(7). Dorsum of abdomen with undivided red spot narrowing anteriad (Figs. 212, 214). Spider found east of Rocky Mountains ...... *descripta* (**Hentz**) (p.114)
   Dorsum of abdomen with large red area divided at midline by black band and covering much of dorsal surface (Fig. 221). Spider found west of Rocky Mountains .... *walsinghami* (**O. Pickard-Cambridge**) (p.117)
9(1). Copulatory openings large, elliptical or slitlike, connected by transverse groove (e.g., Figs. 183, 193, 202). Dorsum of abdomen without red area on posterior half, usually with transverse bands of white setae (Figs. 10, 198) 10
   Copulatory openings small, round, not connected by transverse groove (Figs. 217, 218). Dorsum of abdomen with red area on posterior half, without transverse bands of white setae (Figs. 212, 214, 221) 15
10(9). Femora I and II with dark brown or black longitudinal bands along dorsal, prolateral, and retrolateral surfaces (Fig. 190). Spermathecae little narrower in posterior half than in anterior half (Fig. 198) .... *cingulata* (**C.L. Koch**) (p.105)

Femora I and II without dark longitudinal bands. Spermathecae much narrower in posterior half than in anterior half (e.g., Figs. 185, 194, 205) 11

11(10). Femora I and II (as well as III and IV) orange brown with contrasting yellow or orange tips . . . . *variata* **Gertsch** (p. 111)

Femora I and II without contrasting yellow or orange tips 12

12(11). Dorsum of abdomen with four or five transverse bands of black setae. Copulatory tubes beginning at level of posterior ends of spermathecae (Fig. 209) . . . .
. . . . *alteranda* **Gertsch** (p. 112)

Dorsum of abdomen with transverse bands of white setae (Figs. 10, 198). Copulatory tubes beginning anterior to level of posterior ends of spermathecae (Figs. 185, 194) 13

13(12). Copulatory openings oblique (Fig. 193). Dorsum of abdomen with four or more transverse bands of white setae distributed on both anterior and posterior parts (Fig. 10) . . . . *longipalpa* (**Hentz**) (p. 107)

Copulatory openings essentially transverse (Figs. 183, 196). Dorsum of abdomen with two to four transverse bands of white setae distributed on anterior part (Fig. 198) 14

14(13). Dorsum of abdomen with two transverse bands of white setae. Copulatory tubes directed anteriad (Figs. 185, 186) . . . . *trilineata* (**Hentz**) (p. 104)

Dorsum of abdomen wih three or four transverse white bands, the posterior one interrupted at midline (Fig. 198). Copulatory tubes directed posteriad (Figs. 194, 197) . . . . *gertschi* **Kaston** (p.109)

15(9). Dorsum of abdomen with undivided red spot or spots that narrow anteriad (Figs. 212, 214). Spider found east of Rocky Mountains (Map 33) . . . .
. . . . *descripta* (**Hentz**) (p.114)

Dorsum of abdomen with large red area (often faded in alcohol) divided at midline by black band (Fig. 221). Spider found west of Rocky Mountains (Map 33) . . . . *walsinghami* (**O. Pickard-Cambridge**) (p.117)

Clé des espèces de *Castianeira*

1. Mâle 2

Femelle 9

2(1). Embolus long, étroit, sans grandes dents (p. ex., fig. 187, 211 et 220). Cymbium sans macroseta dorsale 3

Embolus bulbeux, avec grandes dents (fig. 182 et 184). Cymbium avec macroseta dorsale près de la marge rétrolatérale . . . .
. . . . *trilineata* (**Hentz**) (p. 104)

3(2). Fémurs I et II avec bandes longitudinales brun foncé sur les faces dorsale, prolatérale et rétrolatérale (fig. 190) . . . . *cingulata* (**C.L. Koch**) (p. 105)

Fémurs I et II sans bandes longitudinales foncées 4

4(3). Embolus court (longueur de l'embolus ÷ longueur de la tégule + embolus × 100 = 13 ou moins) 5

Embolus plus long (longueur de l'embolus ÷ longueur de la tégule + embolus × 100 = 15 ou plus) 6

5(4). Carapace large de 1,74 à 2,40 mm. Dorsum de l'abdomen avec quatre bandes transversales ou plus de soies blanches distribuées sur les parties antérieure et postérieure (comme dans la figure 10) . . . *longipalpa* (**Hentz**) (p. 107)

Carapace large de 1,20 à 1,60 mm. Dorsum de l'abdomen avec trois ou quatre bandes transversales pâles sur la moitié antérieure (comme dans la figure 198) . . . . *gertschi* **Kaston** (p. 109)

6(4). Fémurs I et II (ainsi que III et IV) brun orangé foncé avec les extrémités jaunes ou orangées contrastantes . . . . *variata* **Gertsch** (p.111)

Fémurs I et II sans extrémités jaunes ou orangées contrastantes 7

7(6). Embolus long (longueur de l'embolus ÷ longueur de la tégule + embolus × 100 = 25 – 26). Dorsum de l'abdomen sans zone rouge postérieure, avec quatre ou cinq bandes transversales de soies noires . . . . *alteranda* **Gertsch** (p. 112)

Embolus moins long (longueur de l'embolus ÷ longueur de la tégule + embolus × 100 = 15 – 17). Dorsum de l'abdomen avec zone rouge postérieure (parfois jaune dans l'alcool), et bandes transversales de soies blanches (fig. 212, 214 et 221) 8

8(7). Dorsum de l'abdomen avec zone rouge entière se rétrécissant à sa partie antérieure (fig. 212 et 214). Araignée rencontrée à l'est des montagnes Rocheuses . . . . *descripta* **(Hentz)** (p.114)

Dorsum de l'abdomen avec large zone rouge divisée à la ligne médiane par une bande noire et recouvrant la majeure partie de la face dorsale (fig. 221). Araignée rencontrée à l'ouest des montagnes Rocheuses . . . . *walsinghami* **(O. Pickard-Cambridge)** (p.117)

9(1). Ouvertures copulatoires grandes, elliptiques ou en forme de fente, réunies par une gouttière transversale (p. ex., fig. 183, 193 et 202). Dorsum de l'abdomen sans zone rouge sur la moitié postérieure, généralement avec bandes transversales de soies blanches (fig. 10 et 198) 10

Ouvertures copulatoires petites, rondes, non réunies par une gouttière transversale (fig. 217 et 218). Dorsum de l'abdomen avec zone rouge sur la moitié postérieure, sans bandes transversales de soies blanches (fig. 212, 214 et 221) 15

10(9). Fémurs I et II avec bandes longitudinales brun foncé ou noires le long des faces dorsale, prolatérale et rétrolatérale (fig. 190). Spermathèques un peu plus étroites sur la moitié postérieure que sur la moitié antérieure (fig. 198) . . . . *cingulata* **(C.L. Koch)** (p. 105)

Fémurs I et II sans bandes longitudinales foncées. Spermathèques beaucoup plus étroites sur la moitié postérieure (p. ex., fig. 185, 194 et 205) 11

11(10). Fémurs I et II (ainsi que III et IV) brun orangé avec extrémités contrastantes jaunes ou orangées . . . . *variata* **Gertsch** (p.111)

Fémurs I et II sans extrémités contrastantes jaunes ou orangées 12

12(11). Dorsum de l'abdomen avec quatre ou cinq bandes transversales de soies noires. Tubes copulatoires commençant au niveau des extrémités postérieures des spermathèques (fig. 209) . . . . *alteranda* **Gertsch** (p.112)

Dorsum de l'abdomen avec bandes transversales de soies blanches (fig. 10 et 198). Tubes copulatoires commençant à l'avant du niveau des extrémités postérieures des spermathèques (fig. 185 et 194) 13

13(12). Ouvertures copulatoires obliques (fig. 193). Dorsum de l'abdomen avec quatre bandes transversales ou plus de soies blanches distribuées sur les parties antérieure et postérieure (fig. 10) . . . . *longipalpa* **(Hentz)** (p.107)

Ouvertures copulatoires essentiellement transversales (fig. 183 et 196). Dorsum de l'abdomen avec de deux à quatre bandes transversales de soies blanches distribuées sur la partie antérieure (fig. 198) 14

14(13). Dorsum de l'abdomen avec deux bandes transversales de soies blanches. Tubes copulatoires dirigés antérieurement (fig. 185 et 186) . . . . *trilineata* **(Hentz)** (p.104)

Dorsum de l'abdomen avec trois ou quatre bandes blanches transversales, la bande postérieure interrompue à la ligne médiane (fig. 198). Tubes copulatoires dirigés postérieurement (fig. 194 et 197) . . . .

Castianeira trilineata (Hentz)

Figs. 182-186; Map 28

Herpyllus trilineatus Hentz, 1847:460, fig. 18 (pl. 24).
Castianeira trilineata: Banks, 1910:11; Reiskind 1969:219, figs. 108 – 110, 119.
Castianeira stupkai Barrows, 1940:137, figs. 10, 10*b*, 10*c*.

Male. Total length approximately 5.40 mm; carapace 2.59 ± 0.10 mm long, 1.55 ± 0.06 mm wide (13 specimens measured). Carapace red orange, darker in eye area, with black margin. Chelicerae orange. Legs yellow orange, sometimes with black smudges along prolateral and retrolateral surfaces, paler distally except tibia and basitarsus IV. Abdomen orange black, with two transverse bands of white scalelike setae on anterior half, with dark shiny scutum covering most of dorsum, and with dark epigastric and ventral scuta. Tibia of palpus slightly longer than wide, concave prolateroventrally, with low retrolateroventral ridge bearing blunt tooth (Fig. 182). Cymbium with dorsal macroseta near retrolateral margin. Tegulum rounded at base, elongate and slender distally; embolus bulbous, toothed (Figs. 182, 184).

Female. Total length approximately 6.50 mm; carapace 2.77 ± 0.15 mm long, 1.73 ± 0.10 mm wide (20 specimens measured). General structure and color essentially as in male but dorsal scutum reduced to small piece at anterior end of

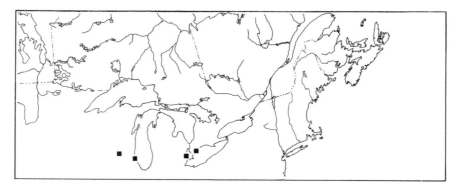

Map 28. Collection localities of *Castianeira trilineata*.

abdomen and rectangular ventral scutum absent. Epigynum with rounded convex plate; copulatory openings large, elliptical, connected by deep transverse groove (Fig. 183). Copulatory tubes short, slender, extending anteromesad, beginning anterior to level of posterior ends of spermathecae; spermathecae broad anteriorly, slender posteriorly, rugose, touching at midline (Figs. 185, 186).

Comments. Males of *C. trilineata* are distinguished from those of the other species by the bulbous toothed embolus. Females of *trilineata* differ from those of the other species by the following combination of characters: dorsum of abdomen with two transverse bands of white setae; leg femora unbanded, without contrasting colors at tips; copulatory openings large, elliptical, connected by deep transverse groove; copulatory tubes beginning anterior to level of posterior ends of spermathecae.

Range. Texas to Florida, northward to Wisconsin and southern Ontario.

Biology. Specimens of *C. trilineata* have been collected in leaf litter among shrubs, in oak forests, and in weedy fields. The Ontario specimens were collected in pitfall traps along the marsh trail in Rondeau Provincial Park and in a relict tall-grass prairie at Windsor. Males were caught from early May to late August, and females from July to September. In general size and color this spider is thought to resemble large workers of the ant *Camponotus castaneus*.

Castianeira cingulata (C.L. Koch)

Figs. 187–191; Map 29

Corinna cingulata C.L. Koch, 1842:22, fig. 706.
Herpyllus zonarius Hentz, 1847:460, fig. 17 (pl. 24).
Castianeira bivittata Keyserling, 1887:442, fig. 16 (pl. 6).
Castianeira cingulata: Simon, 1897:172; Reiskind 1969:221, figs. 100–103, 116–118, frontispiece.
Thargalia canadensis Banks, 1897:194.

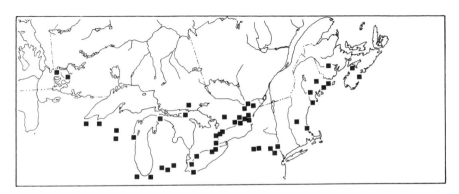

Map 29. Collection localities of *Castianeira cingulata*.

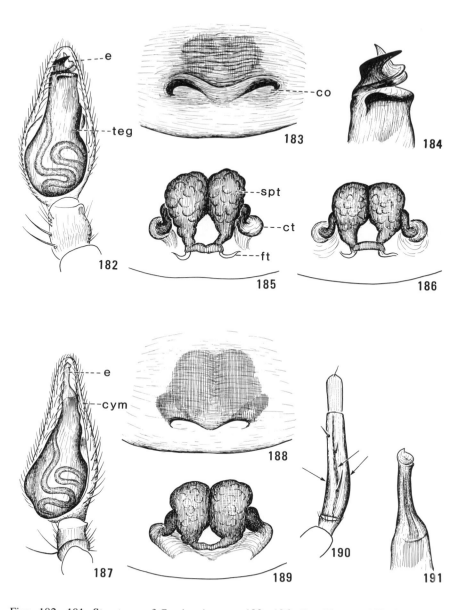

Figs. 182–191. Structures of *Castianeira* spp. 182–186, *C. trilineata*. 182, Palpus of male, ventral view; 183, Epigynum; 184, Embolus of male, ventral view; 185, 186, Spermathecae, dorsal view. 187–191, *C. cingulata*. 187, Palpus of male, ventral view; 188, Epigynum; 189, Spermathecae, dorsal view; 190, femur of leg I, dorsal view; 191, Embolus of male, ventral view. *co*, copulatory opening; *ct*, copulatory tube; *cym*, cymbium; *e*, embolus; *ft*, fertilization tube; *spt*, spermatheca; *teg*, tegulum.

Male. Total length approximately 4.35 mm; carapace 2.12 ± 0.25 mm long, 1.30 ± 0.14 mm wide (15 specimens measured). Carapace dark orange brown anteriorly except eye area which is black; red brown or dark orange posteriorly. Chelicerae orange brown. Legs yellow orange, darker at bases of femora, with indistinct dark longitudinal bands along dorsal, prolateral, and retrolateral surfaces (Fig. 190). Abdomen red black or orange black, with two or three transverse bands of white setae on anterior half, with shiny scutum covering most of dorsum, and with dark epigastric and ventral scuta. Tibia of palpus concave prolateroventrally, with low retrolateroventral ridge bearing small tooth (Fig. 187). Tegulum rounded at base, elongate and slender distally; embolus long, straight, moderately stout, with small spiral at tip (Figs. 187, 191).

Female. Total length approximately 7.25 mm; carapace 3.19 ± 0.26 mm long, 2.04 ± 0.18 mm wide (20 specimens measured). General structure and color essentially as in male but dorsal scutum reduced to small piece at anterior end of abdomen and rectangular ventral scutum absent. Epigynum with rounded convex plate; copulatory openings large, elliptical or slitlike, connected by deep transverse groove (Fig. 188). Copulatory tubes short, broad, extending anterolaterad then anteromesad; spermathecae rounded anteriorly, somewhat narrower posteriorly, rugose, touching at midline (Fig. 189).

Comments. The presence of two or three transverse bands of white setae on the anterior half of the abdomen in *C. cingulata* produces a superficial resemblance to specimens of several other species. Specimens of *C. cingulata* are distinguished from those of the other species by the possession of dark longitudinal bands along the dorsal, prolateral, and retrolateral surfaces of the leg femora.

Range. Ontario to Nova Scotia, southward to Kansas and Florida.

Biology. Specimens of *C. cingulata* have been collected in leaf litter and beneath logs and stones in elm, maple, oak, beech, poplar, and hickory forests. Mature males have been collected from June to September and mature females from May to October and occasionally in winter. The adults resemble worker carpenter ants in structure, color, and locomotion. Egg sacs were found by Kaston (1948) in early April.

Castianeira longipalpa (Hentz)

Figs. 9, 10, 192–195; Map 30

Herpyllus longipalpus Hentz, 1847:457, fig. 9 (pl. 24).
Agroeca tristis Keyserling, 1887:436, fig. 11 (pl. 6).
Geotrecha pinnata Emerton, 1890:170, figs. 4, 4a (pl. 3).
Thargalia perplexa Banks, 1892:15, figs. 53, 53a (pl. 1).
Corinna media Banks, 1896:66.
Corinna pacifica Banks, 1896:66.
Castianeira longipalpis: Banks, 1910:11; Reiskind 1969:186, figs. 7–10, 50–53, frontispiece (*longipalpus*).
Castianeira longipalpa: Bonnet, 1956:967 (footnote 34).

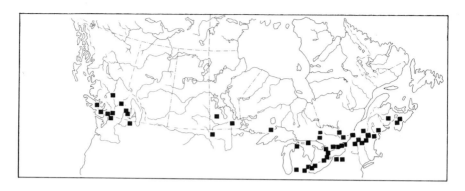

Map 30. Collection localities of *Castianeira longipalpa*.

Male.   Total length approximately 5.65 mm; carapace 2.72 ± 0.16 mm long, 1.89 ± 0.09 mm wide (20 specimens measured). Carapace dark orange brown with purple iridescence, with few darker lines radiating from dorsal groove area, and with eye area black. Chelicerae orange brown. Legs orange brown with purple iridescence, with distal segments yellow orange (except patella, tibia, and basitarsus IV). Abdomen orange brown to purple or nearly black, with four or more transverse bands of white setae distributed on anterior and posterior parts of dorsum (as in Fig. 10), with dark scutum covering anterior three-fourths of dorsum, and with dark epigastric and ventral scuta. Tibia of palpus concave prolateroventrally, with retrolateroventral ridge bearing two teeth (Fig. 192). Tegulum rounded at base, elongate and slender distally; embolus arising at tip of tegulum, short, stout, with tip minutely twisted (Figs. 192, 195).

Female.   Total length approximately 6.90 mm; carapace 3.08 ± 0.13 mm long, 2.03 ± 0.10 mm wide (20 specimens measured). General structure and color essentially as in male but dorsal scutum reduced to small piece at anterior end of abdomen (Fig. 10) and rectangular ventral scutum absent. Epigynum with rounded convex plate; copulatory openings large, elliptical, oblique, connected by transverse groove (Fig. 193). Copulatory tubes short, slender, extending anteromesad, beginning anterior to level of posterior ends of spermathecae; spermathecae rounded anteriorly, elongate and slender posteriorly, rugose, touching at midline (Fig. 194).

Comments.   Specimens of *C. longipalpa* are distinguished from the other species by the following combination of characters: dorsum of abdomen with four or more transverse bands of white setae distributed on both anterior and posterior parts; embolus of male slender but short (length embolus ÷ length tegulum + embolus × 100 = 13 or less); leg femora without dark longitudinal bands or contrasting yellow or orange tips; copulatory tubes of female beginning anterior to level of posterior ends of spermathecae.

Range.   British Columbia to Nova Scotia, southward to Utah, Oklahoma, and Florida.

Biology. Specimens of *C. longipalpa* have been collected from leaf litter in beech−maple and black oak forests, under stones in prairies, by pitfall traps in calcareous and sphagnum bogs, sagebrush rangelands, hayfields, and abandoned cropland, and on buildings. Mature males have been collected from June to October, and mature females from June to November. These spiders are thought to resemble the workers of an unidentified species of mound-building ant. Montgomery (1909) described mating and egg sac construction.

Castianeira gertschi Kaston

Figs. 196−200; Map 31

Castianeira gertschi Kaston, 1945:6, figs. 27−29; Reiskind 1969:217, figs. 104−107, 120.

Male. Total length approximately 4.80 mm; carapace 2.21−2.42 mm long, 1.30−1.49 mm wide (four specimens measured). Carapace dull orange, with narrow dark marginal band, and with several indistinct darker bands radiating from dorsal groove area. Chelicerae dull orange. Legs orange, paler distad except leg IV which becomes darker distad. Abdomen nearly black, with three or four pale transverse bands on anterior half, with posterior band interrupted at midline. Tibia of palpus slightly longer than wide, concave on prolateroventral surface, with retrolateroventral ridge bearing one sharp tooth (Fig. 199). Tegulum rounded at base, elongate and slender distad; embolus arising at tip of elongate part of tegulum, short, dark, stout, with small twist at tip (Figs. 199, 200).

Female. Total length approximately 5.85 mm; carapace 2.36−2.67 mm long, 1.43−1.59 mm wide (eight specimens measured). General structure essentially as in male but dorsal scutum reduced to small piece at anterior end of abdomen (Fig. 198) and ventral scutum absent. Epigynum with rounded convex

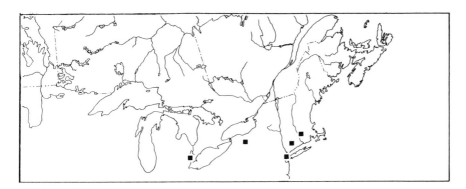

Map 31. Collection localities of *Castianeira gertschi*.

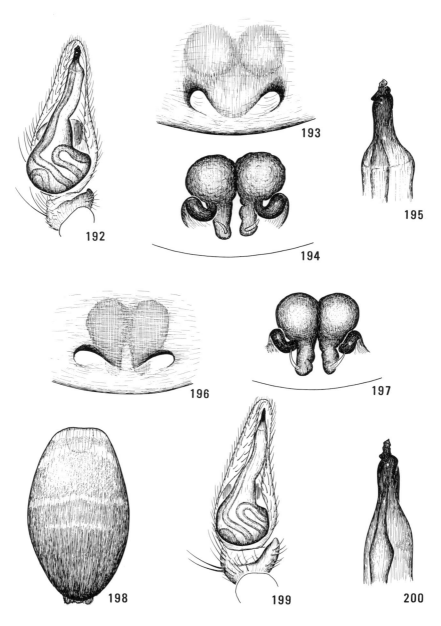

Figs. 192−200. Structures of *Castianeira* spp. 192−195, *C. longipalpa*. 192, Palpus of male, ventral view; 193, Epigynum; 194, Spermathecae, dorsal view; 195, embolus of male, ventral view. 196−200, *C. gertschi*. 196, Epigynum; 197, Spermathecae, dorsal view; 198, Abdomen of female, ventral view; 199, Palpus of male, ventral view; 200, Embolus of male, ventral view.

plate; copulatory openings large, elliptical, connected by curved transverse groove (Fig. 196). Copulatory tubes short, slender, extending mesad, beginning anterior to level of posterior ends of spermathecae; spermathecae rounded anteriorly, elongate and slender posteriorly, rugose, touching at midline (Fig. 197).

Comments.   Specimens of *C. gertschi* are distinguished by the following combination of characters: dorsum of abdomen with three or four pale transverse bands on anterior half (the posterior band interrupted at midline); embolus of male straight but relatively short (length embolus ÷ length tegulum + embolus × 100 = 13 or less); leg femora without dark longitudinal bands or contrasting yellow or orange tips; female copulatory openings large and elliptical; copulatory tubes of female beginning anterior to level of posterior ends of spermathecae.

Range.   Texas to Florida, northward to southern Ontario and Massachusetts.

Biology.   Canadian specimens of *C. gertschi* were collected in pitfall traps in tall grass and under pin oaks in a relict prairie. Mature males are collected from May to July, and mature females from May to August.

Castianeira variata Gertsch

Figs. 201–206; Map 32

Castianeira variata Gertsch, 1942:6, fig. 21; Reiskind 1969:197, figs. 19–23, 60, 61, 78.

Male.   Total length approximately 5.75 mm; carapace 2.58–2.90 mm long, 1.66–1.85 mm wide (four specimens measured). Carapace dark orange brown or dark red brown to nearly black. Chelicerae dark orange brown to nearly black, paler posteriorly. Legs with coxae, trochanters, and distitarsi dull yellow;

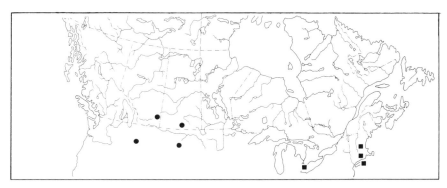

Map 32. Collection localities of *Castianeira variata* (■) and *C. alteranda* (●).

femora dark orange brown to nearly black with contrasting dull yellow or orange tips; tibiae and basitarsi yellow except tibia IV which is nearly black with orange ends and basitarsus IV which is nearly black. Abdomen orange black to black with purple iridescence, with six to eight transverse bands of white setae, with dark scutum covering most of dorsum, and with dark epigastric and ventral scuta. Tibia of palpus approximately as long as wide, concave prolateroventrally, with low ridge retrolateroventrally bearing small tooth (Fig.201). Tegulum rounded at base, elongate and slender distad; embolus arising at tip of elongate part of tegulum, long, dark, slender, with small twist at tip (Figs. 201, 204).

Female. Total length approximately 8.00 mm; carapace 2.95−3.57 mm long, 1.87−2.38 mm wide (three specimens measured). General structure and color essentially as in male but dorsal scutum reduced to small piece at anterior end of abdomen and large ventral scutum absent; dorsum of abdomen with additional transverse white band at posterior end. Epigynum with rounded convex plate; copulatory openings large, elliptical, connected by shallow curved transverse groove (Figs. 202, 203). Copulatory tubes short, rather wide, extending anteriad then posteromesad; spermathecae rounded anteriorly, elongate and slender posteriorly, rugose, touching at midline (Figs. 205, 206).

Comments. Specimens of C. variata are distinguished from those of the other species by the following combination of characters: leg femora dark orange brown without longitudinal bands but with contrasting yellow or orange tips; embolus of male long, slender (length embolus ÷ length tegulum + embolus × 100 = 15 or more); dorsum of abdomen with six to eight transverse bands of white setae; copulatory tubes wide.

Range. Louisiana and Delaware northward to southern Ontario and Massachusetts.

Biology. Specimens of C. variata have been collected by pitfall traps in tall grass and under pin oaks in a relict prairie. Mature males were collected from May to July, and mature females from June to September.

Castianeira alteranda Gertsch

Figs. 207−210; Map 32

Castianeira alteranda Gertsch, 1942:6, figs. 19, 20; Reiskind 1969:206, figs. 66−69, 83.

Figs. 201−210. Genitalia of *Castianeira* spp. 201−206, *C. variata*. 201, Palpus of male, ventral view; 202, 203, Epigynums; 204, Embolus of male, ventral view; 205, 206, Spermathecae, dorsal view. 207−210, *C. alteranda*. 207, Palpus of male, ventral view; 208, Epigynum; 209, Spermathecae, dorsal view; 210, Embolus of male, ventral view.

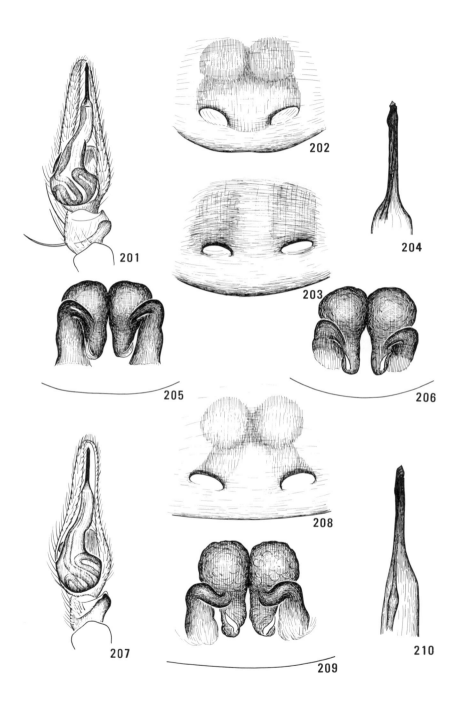

201

202

203

204

205

206

207

208

209

210

113

Male. Total length approximately 6.70 mm; carapace 3.24, 3.27 mm long, 2.09, 2.12 mm wide (two specimens measured). Carapace orange with indistinct darker bands radiating from dorsal groove area, with eye area nearly black. Chelicerae dark orange. Legs orange, except femora I–IV and tibia and basitarsus IV, which are darker. Abdomen dark orange with four or five transverse black bands, with scutum covering most of dorsum, and with epigastric and ventral scuta. Tibia of palpus slightly longer than wide, concave prolateroventrally, with retrolateral ridge bearing one sharp tooth (Fig. 207). Tegulum rounded at base, elongate and slender distad; embolus arising at tip of tegulum, long and slender, with spiral at tip (Figs. 207, 210).

Female. Total length approximately 7.65 mm; carapace 3.54 mm long, 2.34 mm wide (one specimen measured). General structure and color essentially as in male but abdomen with dorsal scutum reduced to small piece at anterior end, and with epigastric scutum indistinct and ventral scutum absent. Epigynum with broad slightly convex plate; copulatory openings moderately large, distinct, connected by shallow groove (Fig. 208). Copulatory tubes short, broad, extending laterad then posteriad, beginning at level of posterior ends of spermathecae; spermathecae rounded anteriorly, more slender posteriorly, rugose, touching at midline (Fig. 209).

Comments. Specimens of *C. alteranda* are distinguished from those of the other species in the genus by the following combination of characters: embolus moderately long, slender (length embolus ÷ length tegulum + embolus × 100 = 25–26); leg femora without dark longitudinal bands, without contrasting pale tips; dorsum of abdomen with four or more black transverse bands; copulatory openings of female large, connected by shallow groove; copulatory tubes beginning at level of posterior ends of spermathecae.

Range. Southern Alberta and Saskatchewan, southward to Colorado.

Biology. The habitat of *C. alteranda* is not recorded. Mature individuals of both sexes have been collected in May.

Castianeira descripta (Hentz)

Figs. 211–217; Map 33

Herpyllus descriptus Hentz, 1847:456, fig. 7 (pl. 24).
Thargalia agilis Banks, 1892:15, figs. 52, 52a (pl. 1).
Thargalia fallax Banks, 1892:16, fig. 54 (pl. 1)
Castianeira descripta: Banks, 1910:11; Reiskind 1969:208, figs. 88–91, 121.

Male. Total length approximately 6.50 mm; carapace 3.03 ± 0.40 mm long, 1.96 ± 0.22 mm wide (20 specimens measured). Carapace dark orange brown, with few indistinct darker lines radiating from dorsal groove area, sometimes with purple iridescence. Chelicerae orange brown. Legs orange brown with purple iridescence (coxae to femora I and II, coxa to tibia III, coxa to

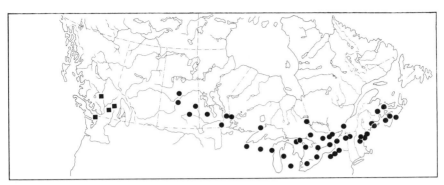

Map 33. Collection localities of *Castianeira descripta* (●) and *C. walsinghami* (■).

basitarsus IV), with distal segments yellow orange. Abdomen orange brown to pale purple or nearly black, with undivided red (fading to dull yellow in alcohol) spot or spots narrowing anteriad, with shiny scutum covering approximately three-fourths of dorsum, and with dark epigastric and ventral scuta. Tibia of palpus slightly longer than wide, concave prolateroventrally, with retrolateroventral ridge bearing two small teeth (Fig. 211). Tegulum rounded at base, elongate and slender distad; embolus arising at tip of tegulum, short, slender, with minute twist at tip (Figs. 211, 213).

Female.   Total length approximately 7.65 mm; carapace 3.20 ± 0.38 mm long, 2.16 ± 0.16 mm wide (20 specimens measured). General structure and color essentially as in male but dorsal sclerite reduced to small piece at anterior end of abdomen (Figs. 212, 214) and large ventral sclerite absent. Epigynum with rounded convex plate; copulatory openings small, round, not connected by transverse groove (Fig. 217). Copulatory tubes short, broad, extending mesad; spermathecae rounded anteriorly, slightly narrowed posteriorly, rugose, touching at midline (Figs. 215, 216).

Comments.   Specimens of *C. descripta* are distinguished from those of the other species by the following combination of characters: leg femora without dark longitudinal bands or contrasting yellow or orange tips; dorsum of abdomen with red spot or spots at posterior end and without transverse white bands; embolus of male relatively short and slender (length embolus ÷ length tegulum + embolus × 100 = 15−17); female copulatory openings small, round, not connected by transverse groove.

Reiskind (1969, p. 208) remarks that *C. descripta* and *C. walsinghami* are "allopatric, closely related, and may . . . be part of one polytypic species, but they are separated, conditionally, on the basis of their pattern differences and slight genitalic differences."

Range.   Saskatchewan to Nova Scotia, southward to Texas and Florida.

Biology. Specimens of *C. descripta* have been collected from leaf litter in dry deciduous forests, under stones and boards or in the open in prairies, and in hayfields, beaches, and sand dunes. Mature males have been collected from June to August, and mature females from July to September. This spider is thought to resemble certain velvet ants in its appearance and jerky movements. Kaston (1948) reports a female with her egg sac under a stone in late August.

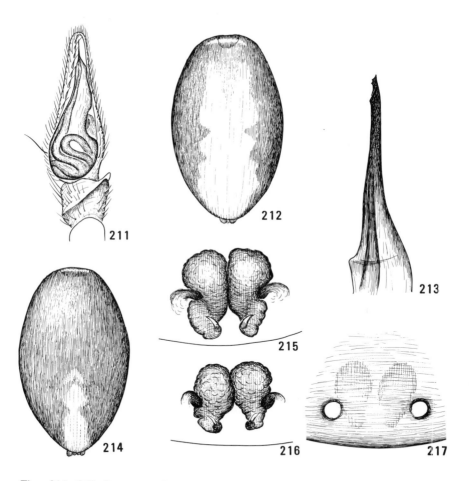

Figs. 211–217. Structures of *Castianeira descripta*. 211, Palpus of male, ventral view; 212, 214, Abdomens of female, dorsal view; 213, Embolus of male, ventral view; 215, 216, Spermathecae, dorsal view; 217, Epigynum.

Figs. 218–222. Structures of *Castianeira walsinghami*. 218, Epigynum; 219, Spermathecae, dorsal view; 220, Palpus of male, ventral view; 221, Abdomen of female, dorsal view; 222, Embolus of male, ventral view.

116

Castianeira walsinghami (O. Pickard-Cambridge)

Figs. 218–222; Map 33

Agroeca walsinghami O. Pickard-Cambridge, 1874:416.
Castianeira walsinghami: Simon, 1897:167; Reiskind 1969:213, figs. 92–95, 122.

Male. Total length approximately 6.45 mm; carapace 3.04 mm long, 2.12 mm wide (one specimen measured). Carapace dark brown to nearly black, darker in eye area. Chelicerae dark brown. Legs with femora I–IV, tibia and basitarsus IV dark brown to black, with other distal segments yellow orange. Abdomen red (usually faded to dull yellow in alcohol), with black median band anteriorly and black indented lateral margins, with shiny scutum covering nearly entire dorsum, and with dark epigastric and ventral scuta. Tibia of palpus slightly longer than wide, concave prolateroventrally, with ridge retrolateroventrally bearing two small teeth (Fig. 220). Tegulum rounded at base, elongate and slender distad; embolus arising at tip of tegulum, short, slender, with minute twist at tip (Figs. 220, 222).

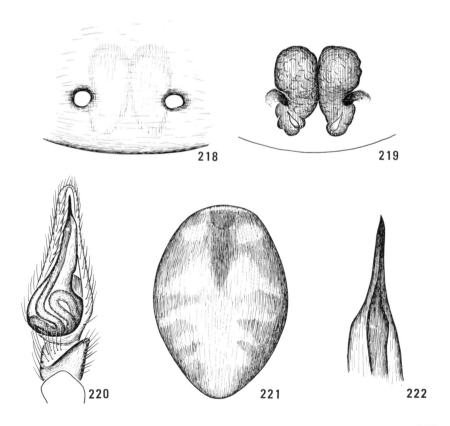

218   219

220   221   222

Female. Total length approximately 7.30 mm; carapace 2.96—3.43 mm long, 2.13—2.36 mm wide (five specimens measured). General structure and color essentially as in male but abdomen with dorsal scutum reduced to small piece at anterior end of abdomen (Fig. 221), and without large ventral sclerite. Epigynum with rounded convex plate; copulatory openings small, round, not connected by transverse groove (Fig. 218). Copulatory tubes short, broad, extending mesad; spermathecae rounded anteriorly, slightly narrowed posteriorly, rugose, touching at midline (Fig. 219).

Comments. Specimens of *C. walsinghami* are distinguished by the following combination of characters: leg femora without dark longitudinal bands or contrasting yellow or orange tips; dorsum of abdomen with large divided red area (usually faded to dull yellow in alcohol) and without transverse white bands; embolus of male relatively short and slender (length embolus ÷ length tegulum + embolus × 100 = 15—17); female copulatory openings small, round, not connected by transverse groove.

The relationship between *walsinghami* and *C. descripta* is uncertain as noted in the comments on *descripta*.

Range. British Columbia southward to Oregon.

Biology. Worley's (1932) specimens of *C. walsinghami* were found beneath stones in a conifer forest. Females have been collected from late July to early September.

Genus *Agroeca* Westring

Spiders of the genus *Agroeca* are of medium size (body length approximately 5.50 mm) with compact bodies and legs and with small eyes set in a tight group at the anterior margin of the carapace. They are covered with short inconspicuous setae, and the main color is brown or orange mottled with black, gray, dull yellow, and dull red. These cryptic predators inhabit moist litter in fields, meadows, bogs, and shady forests. If disturbed they are swift to regain cover. The most effective collecting methods are pitfall trapping and litter sifting.

Very little is known about the courtship, oviposition, or life history of these spiders, though observations on European species such as *A. brunnea* (Blackwall), which closely resembles *A. ornata* Banks of North America in structure, indicate that courtship probably takes place in the open and that the ovoid egg sac, encased in mud, is hung by a short thread from a low-growing plant. Similar egg sacs have been found in Ontario, covered with mud or with fine particles of plant litter and suspended from the undersides of stones, but without the parent spider to afford identification. Kaston (1948) indicates that in *A. minuta* Banks, however, the egg sac is a planoconvex structure and is attended by the female.

Description. Total length 4.75—6.30 mm. Carapace (Fig. 11) ovoid, longer than wide, rather low, highest at dorsal groove, covered with short recumbent setae; surface orange, with indistinct black bands radiating from dorsal groove area; dorsal groove deep, distinct. Eyes in two transverse rows, uniform in size and spacing; anterior row (dorsal view) straight; posterior row procurved, approximately equal in length to anterior row. Chelicerae orange brown, moderately short, stout; promargin of fang furrow with three teeth; retromargin with two teeth. Palp-coxal lobes slightly wider than long, convex on lateral margin. Legs (Fig. 11) moderately long and strong, hairy, orange, without bands or rings; with thin claw tufts and scopulae; femur I with three dorsal macrosetae, one or two prolaterals; basitarsus I with three pairs of ventral macrosetae, without median unpaired ventral macroseta at tip; trochanter IV with deep notch at tip on ventral side. Abdomen (Fig. 11) elongate ovoid, mottled with black, dull yellow, and dull red; with short erect setae and with cluster of long erect curved setae at anterior end; with inconspicuous narrow scutum on anterior part of dorsum in males. Femur and patella of male palpus without apophyses; tibia approximately twice as long as wide, with tapered pointed retrolateral apophysis, without ventral apophysis (Figs. 223, 228); tegulum convex, with hooked apophysis arising from membranous area near retrolateral margin; embolus pointed or flattened, arising at middle of tegulum or farther distad. Epigynum of female with median septum; copulatory openings inconspicuous, well-separated, located in anterior half of plate (Figs. 224, 226, 229). Copulatory tubes short, inconspicuous, extending posteriad; spermathecae long, sinuous, extending posteriad to level of genital groove, with prominent spermathecal organ (Figs. 225, 227, 230).

Comments.   The palp-coxal lobes, which are convex along the lateral margins in specimens of *Agroeca* spp., distinguish these spiders from specimens of *Cheiracanthium* spp., *Clubiona* spp., and *Clubionoides* spp. The distinctly procurved posterior row of eyes, the trochanteral notch, and the presence of three pairs of ventral macrosetae on basitarsus I distinguish these spiders from specimens of *Trachelas* spp., *Phrurotimpus* spp., and *Scotinella* spp. The presence of three pairs (rather than two pairs) of ventral macrosetae on basitarsus I, the possession of a dense cluster of long erect setae at the anterior end of the abdomen, and the less slender and less antlike body and legs distinguish specimens of *Agroeca* spp. from those of *Castianeira* spp.

The genus *Agroeca* includes a world fauna of approximately 23 species, seven of which occur in North America. Two are represented in Canada. Kaston (1938b) revised the North American species.

Key to species of *Agroeca*

1.   Male 2

        Female 3

2(1).   Embolus slender, angled (Fig. 223) . . . . *pratensis* **Emerton** (p.120)

        Embolus broad, curved (Fig. 228) . . . . *ornata* **Banks** (p.121)

3(1).   Epigynum with median septum short, broad, restricted to anterior part of plate (Fig. 224). Spermathecae looped far laterad (Fig. 225) . . . .
      . . . . *pratensis* **Emerton** (p.120)

        Epigynum with median septum long, slender, extending posteriad to genital groove (Figs. 226, 229). Spermathecae not looped laterad (Figs. 227, 230)
      . . . . *ornata* **Banks** (p.121)

Clé des espèces d'*Agroeca*

1. Mâle 2
    Femelle 3
2(1). Embolus étroit, angulaire (fig. 223) . . . . ***pratensis* Emerton** (p. 120)
    Embolus large, courbé (fig. 228) . . . . ***ornata* Banks** (p. 121)
3(1). Épigyne avec septum médian court, large, limité à la partie antérieure de la plaque
    (fig. 224). Spermathèques enroulées loin latéralement (fig. 225) . . . .
    . . . . ***pratensis* Emerton** (p. 120)
    Épigyne avec septum médian long, étroit, se prolongeant postérieurement
    jusqu'à la gouttière génitale (fig. 226 et 229). Spermathèques non enroulées
    latéralement (fig. 227 et 230) . . . . ***ornata* Banks** (p. 121)

Agroeca pratensis Emerton

Figs. 223–225; Map 34

Agroeca pratensis Emerton, 1890:190, figs. 7–7e (pl. 6); Kaston 1938b:566, figs. 4, 9, 13, 18.

Male. Total length approximately 5.05 mm; carapace 2.41 ± 0.14 mm long, 1.85 ± 0.12 mm wide (19 specimens measured). Carapace dark orange, with black margin and with indistinct black bands radiating from dorsal groove area. Chelicerae orange brown. Legs orange, unmarked. Abdomen mottled with black, dull yellow, and dull red, with paired indistinct black spots along midline of dorsum; with inconspicuous narrow scutum on anterior two-thirds of dorsum. Tibia of palpus approximately twice as long as wide, with tapered pointed retrolateral apophysis, without ventral apophysis (Fig. 223). Tegulum convex, with hooklike apophysis attached to membranous area near retrolateral margin; embolus slender, angled, arising near distal end of tegulum, curved and slender toward tip (Fig. 223).

Female. Total length approximately 6.30 mm; carapace 2.63 ± 0.19 mm long, 1.94 ± 0.12 mm wide (20 specimens measured). General structure and color

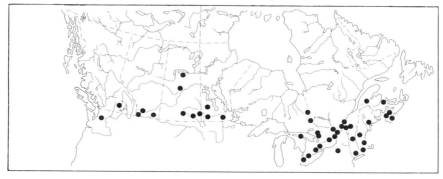

Map 34. Collection localities of *Agroeca pratensis*.

essentially as in male but abdomen lacking dorsal scutum. Epigynum with elongate plate; median septum short, broad, convex, restricted to anterior end of epigynal plate; copulatory openings inconspicuous, located at lateral margins of median septum (Fig. 224). Copulatory tubes short, inconspicuous, extending posteriad; spermathecae long, looped laterad, curved dorsad near posterior end, each with prominent spermathecal organ (Fig. 225).

Comments.   Specimens of *A. pratensis* closely resemble those of *A. ornata* in size and color but can be distinguished by the slender angular embolus of the male and by the short broad median septum and looped spermathecae of the female.

Range.   British Columbia to Nova Scotia, southward to Utah and to Georgia.

Biology.   Specimens of *A. pratensis* have been collected from the ground in pastures, meadows, wheat fields, shortgrass prairies, sagebrush rangelands, calcareous and sphagnum bogs, marshes, and pin oak and red oak forests. One specimen was collected in a house in late autumn. Mature males have been found from late August to November, and mature females from March to November.

Agroeca ornata Banks

Figs. 11, 226−230; Map 35

Agroeca ornata Banks, 1892:23, figs. 68, 68*a* (pl. 1); Kaston 1938*b*:564, figs. 3, 8, 12, 17.
Agroeca repens Emerton, 1894:412, figs. 6, 6*a* (pl. 2).
Rachodrassus monroensis Kaston, 1938*a*:173, figs. 1−4 (pl. 8).

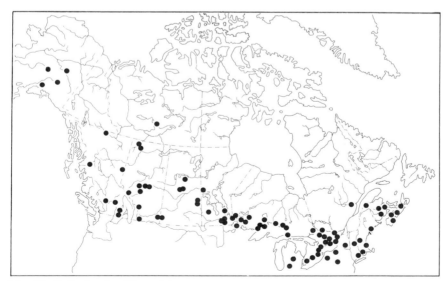

Map 35. Collection localities of *Agroeca ornata*.

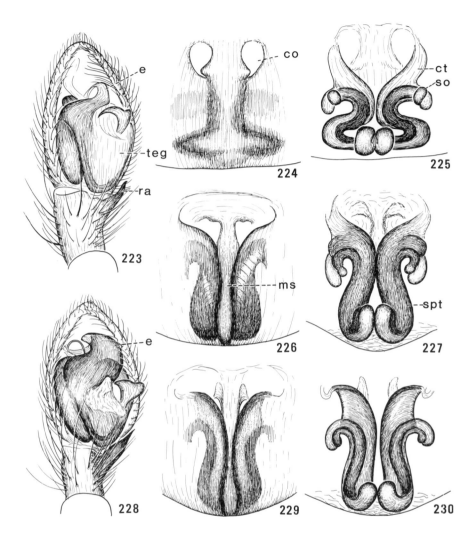

Figs. 223−230. Genitalia of *Agroeca* spp. 223−225, *A. pratensis*. 223, Palpus of male, ventral view; 224, Epigynum; 225, Spermathecae, dorsal view. 226−230, *A. ornata*. 226, 229, Epigynums; 227, 230, Spermathecae, dorsal view; 228, Palpus of male, ventral view. *co*, copulatory opening; *ct*, copulatory tube; *e*, embolus; *ms*, median septum; *ra*, retrolateral apophysis; *so*, spermathecal organ; *spt*, spermatheca; *teg*, tegulum.

Male.   Total length approximately 4.75 mm; carapace 2.28 ± 0.08 mm long, 1.75 ± 0.06 mm wide (20 specimens measured). Carapace dark orange, with indistinct darker bands radiating from dorsal groove area. Chelicerae orange brown. Legs orange, without darker markings. Abdomen mottled with black, dull yellow, and dull red, with indistinct paired black spots along midline; with inconspicuous narrow scutum on anterior two-thirds of dorsum. Tibia of palpus approximately twice as long as wide, with tapered pointed retrolateral apophysis, without ventral apophysis (Fig. 228). Tegulum convex, with hooked apophysis arising from membranous area near retrolateral margin; embolus broad, curved, arising near middle of tegulum, extending to tip of tegulum, curled upon itself, with fine curved tip (Fig. 228).

Female.   Total length approximately 5.55 mm; carapace 2.21 ± 0.26 mm long, 1.68 ± 0.20 mm wide (20 specimens measured). General structure and color (Fig. 11) essentially as in male but abdomen without dorsal scutum. Epigynum with elongate plate; median septum long, slender, extending from anterior end of plate to genital groove; copulatory openings inconspicuous, well-separated, located at sides of median septum (Figs. 226, 229). Copulatory tubes short, inconspicuous; spermathecae elongate, sinuous, lying close together near midline, arched dorsad at posterior ends, not looped laterad, with prominent spermathecal organ (Figs. 227, 230).

Comments.   Specimens of *A. ornata* closely resemble those of *A. pratensis* in size and color but can be distinguished from the latter by the broad curved embolus of the male and by the long slender median septum and straighter spermathecae of the female.

Range.   Alaska to Nova Scotia, southward to California and to New Jersey.

Biology.   The habitats recorded for *A. ornata* are ground litter or decaying logs in fir, pine, spruce, cedar, oak, maple, beech, and poplar forests, and on the ground in pastures, meadows, marshes, sphagnum bogs, mosses, and lichens. Males have been collected from March to November, and females from April to November.

Genus *Trachelas* L. Koch

Spiders of the genus *Trachelas* characteristically have shiny red carapaces and sterna that contrast strikingly with their pale abdomens. They spend considerable time in silk retreats spun in rolled leaves, under loose bark, or on objects on the ground, and are sometimes collected by sweeping or beating. Maturity is apparently attained in autumn, when individuals may be seen in houses, and the eggs are laid in white sacs during the same season. Some have been known to bite when handled; the venom causes local swellings and lesions with severe pain.

Description.   Total length 3.25 − 7.70 mm. Carapace (Fig. 13) dark red to red brown, ovoid in dorsal view, somewhat truncate at posterior margin, highest between dorsal groove and posterior row of eyes, sparsely covered with short pale erect setae that arise from minute tubercles; dorsal groove shallow, usually distinct, located on posterior slope of carapace. Eyes moderately large, uniform in size, arranged in two transverse rows; anterior row slightly recurved (dorsal view), with anterior median eyes slightly closer to anterior laterals than to each other; posterior row recurved, with eyes uniformly spaced or with posterior medians slightly closer to posterior laterals than to each other, slightly longer than anterior row. Chelicerae dark orange, moderately long, stout, hairy, covered with minute tubercles; promargin of fang furrow with three teeth, retromargin with two. Palp-coxal lobes longer than wide, convex on lateral margin. Legs (Fig. 13) yellow, orange, or orange red, long and rather slender, with short dark dense claw tufts and sparse scopulae; without macrosetae; tibiae and tarsi I and II of males (and of females of *T. tranquillus*) with rows of small cusps; trochanter IV without ventral notch at distal end. Abdomen (Fig. 13) pale orange yellow, yellow, or off-white, ovoid, with or without small dorsal scutum at anterior end, covered with short pale semi-erect setae; without cluster of long erect setae at anterior end. Tibia of male palpus as long as wide or slightly longer, with single small apophysis (Figs. 231, 236, 238); tegulum convex, without apophysis; embolus short, erect, tapered, coiled or straight, situated at tip of tegulum (Figs. 231, 236, 238). Epigynum of female with rounded, flat, or convex plate, with or without small hood; copulatory openings round or funnellike (Figs. 232, 234, 237). Copulatory tubes and spermathecae terminating in two branches of different lengths and arising at posterior end of epigynum, with or without complex mesal coils (Figs. 233, 235, 239).

Comments.   Specimens of *Trachelas* spp. lack the lateral constrictions in the palp-coxal lobes found in representatives of *Cheiracanthium* spp., *Clubiona* spp., and *Clubionoides* spp. They lack the procurved posterior eye row and trochanteral notch found in representatives of *Castianeira* spp. and *Agroeca* spp. They have the posterior row of eyes recurved rather than straight, and lack the leg macrosetae and ventral prominence on the male palpal femur found in members of *Phrurotimpus* spp. and *Scotinella* spp.

The genus *Trachelas* comprises a world fauna of approximately 83 species. Twenty-three of these are known from North America, and three occur or are assumed to occur in Canada (Platnick and Shadab 1974*a*, 1974*b*).

Key to species of *Trachelas*

1.   Male 2
        Female 4
2(1).   Carapace width greater than 1.90 mm. Embolus stout, coiled (Fig. 231) . . . . .
        . . . . *tranquillus* (**Hentz**) (p. 125)
        Carapace width less than 1.90 mm. Embolus slender, not coiled (Figs. 236, 238) .
        3
3(2).   Palpal tibia with apophysis short, blunt, not extending to tip of tibia (Fig. 236)
        . . . . *deceptus* (**Banks**) (p. 127)
        Palpal tibia with apophysis long, pointed, extending beyond tip of tibia (Fig. 238)
        . . . . *californicus* **Banks** (p. 129)

124

4(1). Copulatory openings large, situated at middle of epigynal plate (Fig. 232). Copulatory tube and spermatheca coiled (Fig. 233) .... *tranquillus* **(Hentz)** (p. 125) Copulatory openings small, situated in posterior third of epigynal plate (Figs. 234, 237). Copulatory tube and spermatheca not coiled (Figs. 235, 239) 5

5(4). Epigynum without hood; copulatory openings round, conspicuous (Fig. 234). Copulatory tube and spermatheca with mesal branch kidney-shaped and much longer than lateral branch (Fig. 235) ... *deceptus* **(Banks)** (p. 127) Epigynum with hood; copulatory openings funnel-shaped, inconspicuous (Fig. 237). Copulatory tube and spermatheca with mesal branch slender, sinuous, slightly longer than lateral branch (Fig. 239)...................... .... *californicus* **Banks** (p. 129)

Clé des espèces de *Trachelas*

1. Mâle .... 2 Femelle 4

2(1). Carapace plus large que 1,90 mm. Embolus trapu, enroulé (fig. 231) .... *tranquillus* **(Hentz)** (p. 125) Carapace moins large que 1,90 mm. Embolus étroit, non enroulé (fig. 236 et 238) 3

3(2). Tibia palpal avec apophyse courte, obtuse, ne se prolongeant pas jusqu'à l'extrémité du tibia (fig. 236) .... *deceptus* **(Banks)** (p.127) Tibia palpal avec apophyse longue, pointue, se prolongeant au delà de l'extrémité du tibia (fig. 238) .... *californicus* **(Banks)** (p.129)

4(1). Ouvertures copulatoires grandes, situées au milieu de la plaque épigynale (fig. 232). Tube copulatoire et spermathèque enroulés (fig. 233) .... *tranquillus* **(Hentz)** (p. 125) Ouvertures copulatoires petites, situées sur le tiers postérieur de la plaque épigynale (fig. 234 et 237). Tube copulatoire et spermathèque non enroulés (fig. 235 et 239) 5

5(4). Épigyne sans capuchon; ouvertures copulatoires rondes, apparentes (fig. 234). Tube copulatoire et spermathèque avec branche mésale réniforme et beaucoup plus longue que la branche latérale (fig. 235) .... *deceptus* **(Banks)** (p.127) Épigyne avec capuchon; ouvertures copulatoires infundibuliformes, non apparentes (fig. 237). Tube copulatoire et spermathèque avec branche mésale étroite, sinueuse, légèrement plus longue que la branche latérale (fig. 239) .... *californicus* **Banks** (p.129)

Trachelas tranquillus (Hentz)

Figs. 13, 231−233; Map 36

Clubiona tranquilla Hentz, 1847:450; fig. 1 (pl. 30).
Trachelas ruber Keyserling, 1887:439, fig. 14 (pl. 6).
Trachelas tranquillus: Banks, 1891:84; Platnick & Shadab 1974a:8, figs. 1−9, 42−44.

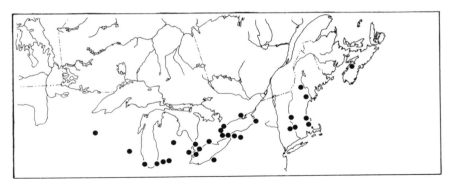

Map 36. Collection localities of *Trachelas tranquillus*.

Male.   Total length approximately 5.55 mm; carapace 2.69 ± 0.22 mm long, 2.22 ± 0.18 mm wide (10 specimens measured). Carapace dark red or red brown, with indistinct black lines radiating from dorsal groove area. Chelicerae dark orange. Legs yellow to pale orange or orange red; tibiae and tarsi I and II with ventral series of minute cusps. Abdomen pale yellow to off-white, with indistinct scutum covering much of dorsum and with thin epigastric scutum. Tibia of palpus with fingerlike retrolateral apophysis (Fig. 231). Tegulum long, convex, without apophysis; embolus stout, tapered, coiled at base, slightly hooked at tip (Fig. 231).

Female.   Total length approximately 7.70 mm; carapace 3.24 ± 0.22 mm long, 2.72 ± 0.17 mm wide (10 specimens measured). General structure and color essentially as in male but abdomen lacking dorsal scutum. Epigynum with rounded flat plate; copulatory openings large, round, dark, situated near middle of epigynal plate (Fig. 232). Copulatory tubes and spermathecae long, slender, with complex coil mesally, giving rise terminally to two branches, the mesal branch longer than the lateral branch and bladderlike (Fig. 233).

Comments.   Specimens of *T. tranquillus* are larger than those of *T. deceptus* and *T. californicus*. Males differ by having a stout coiled embolus, and females differ by having large copulatory openings and coiled copulatory tubes and spermathecae.

Range.   Minnesota to Nova Scotia, southward to Oklahoma and to northern Georgia.

Biology.   Specimens of *T. tranquillus* have been collected by sweep nets and beating trays from the foliage of deciduous trees and shrubs, from the bases of herbs, or from silk retreats inside rolled leaves. Some were found under stones and in the folds and crevices of Malaise traps or other objects on the ground, and many were taken wandering inside houses, particularly in autumn. Mature males have been collected form mid-June to early November, and mature females in all months except February and March (Platnick and Shadab 1974*a*). Both sexes have been known to bite humans, on several occasions causing local swelling and pain, sometimes with ulcer formation (Platnick and Shadab 1974*a*).

Trachelas deceptus (Banks)

Figs. 234–236; Map 37

Meriola decepta Banks, 1895:81.
Trachelas parvulus Banks, 1898:225, fig. 28 (pl. 13).
Meriola inornata Banks, 1901:574, fig. 6.
Meriola decepta floridana Chamberlin & Ivie, 1935:40, fig. 107 (pl. 13).
Trachelas deceptus: Simon, 1897:180; Platnick & Shadab 1974*b*:29, figs.
39, 103–106.

Male.   Total length approximately 3.60 mm; carapace 1.73 ± 0.19 mm long, 1.35 ± 0.18 mm wide (10 specimens measured). Carapace dark red or orange red, with few indistinct black lines radiating from dorsal groove area. Chelicerae dark orange. Legs orange yellow to dark orange; tibiae and tarsi I and II with series of minute cusps on ventral surface. Abdomen pale yellow or off-white with indistinct chevrons and middle and lateral longitudinal bands, without scuta. Tibia of palpus slightly longer than wide, with small blunt apophysis that does not extend to tip of tibia (Fig. 236). Tegulum somewhat convex, without apophysis; embolus minute, pointed, arising smoothly from tip of tegulum (Fig. 236).

Female.   Total length approximately 3.80 mm; carapace 1.61 ± 0.16 mm long, 1.28 ± 0.09 mm wide (10 specimens measured). General structure and color essentially as in male but color usually much darker and legs I and II lacking cusps. Epigynum with rounded slightly convex plate; plate with shallow depression mesally and with low transverse ridge along posterior margin; copulatory openings small, round, distinct, situated in posterior third of epigynal plate (Fig. 234). Copulatory tubes and spermathecae not coiled, arising posteriorly, each comprising two branches, with mesal branch kidney-shaped, much longer than lateral branch (Fig. 235).

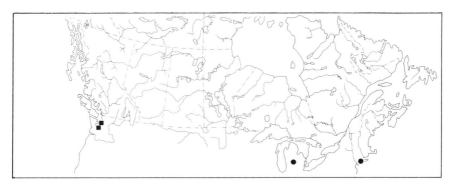

Map 37. Collection localities of *Trachelas deceptus* (●) and *T. californicus* (■).

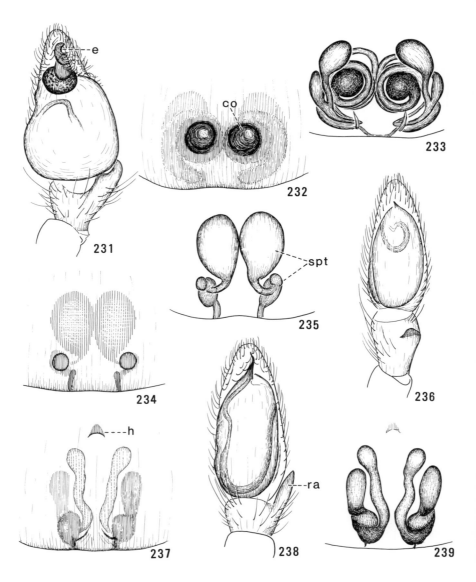

Figs. 231–239. Genitalia of *Trachelas* spp. 231–233, *T. tranquillus*. 231, Palpus of male, ventral view; 232, Epigynum; 233, Spermathecae, dorsal view. 234–236, *T. deceptus*. 234, Epigynum; 235, Spermathecae, dorsal view; 236, Palpus of male, ventral view. 237–239, *T. californicus*. 237, Epigynum; 238, Palpus of male, ventral view; 239, Spermathecae, dorsal view. *co*, copulatory opening; *e*, embolus; *h*, hood; *ra*, retrolateral apophysis; *spt*, spermathecae.

Comments. Specimens of *T. deceptus* are distinguished from those of *T. tranquillus* by their smaller size and by the minute embolus and short retrolateral apophysis on the palpal tibia of males and by the smaller copulatory openings and simpler copulatory tubes and spermathecae of females. Males of *deceptus* differ from those of *T. californicus* by the short retrolateral apophysis on the palpal tibia, and females differ by the conspicuous round copulatory openings and longer, kidney-shaped mesal branch of the copulatory tube and spermatheca.

Range. Guatemala northward to Utah, Michigan, and New York.

Biology. Specimens of *T. deceptus* have been collected from soybean fields and meadows by beating nets and litter sifters, and from plants in greenhouses. Mature males were collected in July, and females from July to October.

Trachelas californicus Banks

Figs. 237–239; Map 37

Trachelas californica Banks, 1904:339, fig. 47 (pl. 40).
Trachelas californicus: Platnick & Shadab, 1974b:32, figs. 40, 107–110.

Male. Total length approximately 3.25 mm; carapace 1.59 ± 0.21 mm long, 1.33 ± 0.16 mm wide (10 specimens measured). Carapace dark orange, with indistinct darker lines radiating from dorsal groove area. Chelicerae dark orange. Legs yellow to pale orange; tibiae and tarsi I and II with ventral series of minute cusps. Abdomen orange yellow, with indistinct scutum, covering part of dorsum. Tibia of palpus approximately as long as wide, with stout tapered retrolateral apophysis extending beyond tip of tibia (Fig. 238). Tegulum convex, without apophysis; embolus short, tapered, sinuous, arising smoothly from distal end of tegulum (Fig. 238).

Female. Total length approximately 4.20 mm; carapace 1.65 ± 0.10 mm long, 1.50 ± 0.17 mm wide (10 specimens measured). General structure and color essentially as in male but abdomen lacking scutum and legs lacking cusps. Epigynum with rounded slightly convex plate; plate shallowly depressed mesally, with low transverse ridge along posterior margin, and with small hood; copulatory openings small, funnellike, inconspicuous, situated near posterior margin of epigynal plate (Fig. 237). Copulatory tubes and spermathecae without mesal coil, with two branches that arise posteriorly and extend anteriad; mesal branch slender, sinuous, slightly longer than lateral branch (Fig. 239).

Comments. Specimens of *T. californicus* are distinguished from those of *T. tranquillus* by their smaller size, by the shorter embolus in males, and by the small copulatory openings and lack of a median coil in the copulatory tubes and spermathecae of females. Males of *californicus* differ from those of *T. deceptus* by the long pointed retrolateral apophysis on the palpal tibia and by the larger embolus, and females differ by the inconspicuous funnellike copulatory openings and slender sinuous mesal branch of the copulatory tubes and spermathecae.

Range. Baja California, northward to Washington.

Biology. Nothing is recorded.

Genus *Phrurotimpus* Chamberlin & Ivie

Spiders of the genus *Phrurotimpus* are small somewhat antlike inhabitants of the litter layer in fields, swamps, and forests. They sometimes make sudden sorties across open rock faces or over fallen leaves, and most of their activity appears to take place during daylight hours (Dondale et al. 1972). If caught in the open, they may suddenly become motionless with the legs flexed above the cephalothorax, thus concealing the striped iridescent carapace and aiding in concealment. No silk retreat is made, and the ovipositing female fastens her flat, scalelike, shiny red egg sacs to stones and then abandons them. The tibiae and basitarsi of leg I are often armed with many pairs of long pale overlapping macrosetea.

Description. Total length 1.75–3.50 mm. Carapace rather low, highest at anterior end of dorsal groove; orange with black margins and network of black lines or longitudinal bands, often sparsely covered with iridescent scalelike setae (Fig. 244); dorsal groove deep, usually distinct. Eyes prominent, arranged in two uniform transverse rows of approximately the same length; anterior row straight or slightly recurved (dorsal view); posterior row straight or slightly procurved; posterior median eyes often angular or ovoid in shape. Chelicerae orange or yellow, sometimes marked with black; thick at base, tapered toward tips; promargin of fang furrow with three well-spaced teeth, and retromargin with three clustered teeth. Palp-coxal lobes approximately as wide as long, convex along lateral margin. Legs orange or yellow marked with black; rather short, slender, with sparse claw tufts, without scopulae; femur I with one dorsal macroseta, two prolaterals; tibia I often swollen and darkened (Figs. 245, 248) with five to seven pairs of long stout overlapping ventral macrosetae; basitarsus I with four pairs of ventral macrosetae (Fig. 245); trochanter IV without ventral notch at distal end. Abdomen rotund, with pattern of dark chevrons in females (Fig. 244), usually covered with scalelike iridescent setae, with large shiny dorsal and epigastric scuta in males, without cluster of long erect setae at anterior end. Femur of male palpus with rounded hairy prominence on ventral surface (e.g., Fig. 240); patella of palpus without apophysis; tibia of palpus slightly longer than wide, with small ventral prominence and with long single stout retrolateral apophysis (e.g., Fig. 240); tegulum convex, with small spurlike apophysis; embolus short, blunt, arising broadly at tip of tegulum (e.g., Fig. 240). Epigynum of female with convex hairy plate (e.g., Figs. 241, 247); copulatory openings conspicuous, situated in anterior half or at midlength of epigynal plate. Copulatory tubes short, funnellike; spermathecae in two parts, with anterior part large, curled and with posterior part smaller, ovoid or angular in outline, lying at genital groove (e.g., Figs. 242, 246).

Comments. Specimens of *Phrurotimpus* spp. are distinguished from those of *Cheiracanthium* spp., *Clubiona* spp., and *Clubionoides* spp. by the lack of a constriction on the lateral margins of the palp-coxal lobes. They differ from specimens of *Castianeira* spp., *Agroeca* spp., and *Trachelas* spp. by having the anterior row of eyes straight or nearly so and by having four pairs of ventral macrosetae on basitarsus I. They differ from specimens of *Scotinella* spp. by having a dorsal macroseta on femur I, a smooth prominence on the ventral surface of the male palpal femur, an unbranched retrolateral apophysis on the male palpal tibia, and conspicuous copulatory openings and two-part spermathecae in females.

This North American genus comprises approximately 23 species, of which five are represented or assumed to be represented in Canada. The genus needs revising.

Key to species of *Phrurotimpus*

(Females of *P. minutus* are not included)

Clé des espèces de *Phrutotimpus*

(Femelles de *P. minutus* non incluses)

Phrurotimpus alarius (Hentz)

Figs. 240–242; Map 38

Herpyllus alarius Hentz, 1847:461, fig. 20 (pl. 24).
Phrurolithus palustris Banks, 1892:23, fig. 70 (pl. 1).
Phrurotimpus alarius: Chamberlin & Ivie, 1935:34; Kaston 1945:5, figs. 38, 39.

Male. Total length approximately 2.20 mm; carapace 1.06 ± 0.05 mm long, 0.86 ± 0.04 mm wide (20 specimens measured). Carapace orange with black lateral margins and with pair of broken lateral black longitudinal bands; lateral areas sometimes with black branching lines. Chelicerae orange. Legs orange, with patellae and basal two-thirds of tibiae I gray; tibiae and basitarsi II–IV with indistinct gray ring near middle; tibia I without ventral fringe of black setae, with five or six pairs of long ventral macrosetae. Abdomen pale with indistinct gray chevrons, and with scutum covering most of dorsum; venter with gray streaks and spots, with epigastric scutum. Femur of palpus with smooth hairy prominence on ventral surface (Fig. 240). Tibia of palpus slightly longer than wide, with small ventral prominence and with long curved retrolateral apophysis;

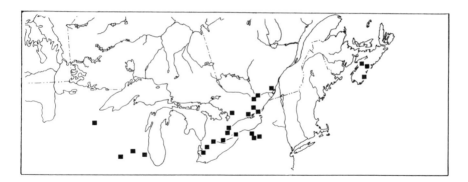

Map 38. Collection localities of *Phrurotimpus alarius*.

retrolateral apophysis nearly as broad at base as width of tibia, smoothly tapered toward tip (Fig. 240). Tegulum convex, with shallow excavation at distal end and with small spurlike apophysis; embolus short, blunt, arising broadly at tip of tegulum (Fig. 240).

Female. Total length approximately 2.90 mm; carapace 1.13 ± 0.05 mm long, 0.93 ± 0.05 mm wide (20 specimens measured). General structure and color essentially as in male but tibia I sometimes with seven (rather than five or six) pairs of long ventral macrosetae, and abdomen lacking scuta. Epigynum with convex plate having convex posterior margin; copulatory openings large, deep, well-separated, situated anterior to spermathecae (Fig. 241). Copulatory tubes short, broad, tapered; spermathecae in two parts, with anterior part large, rounded, and with posterior part small, ovoid, situated at genital groove (Fig. 242).

Comments. Males of *P. alarius* are distinguished from those of the other species by the basally thickened, curved, and smoothly tapered retrolateral apophysis on the palpal tibia. Females of *alarius* are distinguished from those of the other species by the anterior position of the copulatory openings.

Range. New Mexico to Florida, northward to Wisconsin and Nova Scotia.

Biology. Most specimens of *P. alarius* have been collected by pitfall traps in the leaf litter of beech—maple, oak, or birch forests, or in fields, meadows, and marshes. A few were taken under stones, in beach debris, or under piles of lumber or firewood. Mature males and females were collected from May to September.

Phrurotimpus borealis (Emerton)

Figs. 243—248; Map 39

Phrurolithus borealis Emerton, 1911:404, figs. 3, 3*a* (pl. 6).
Phrurolithus utus Chamberlin & Ivie, 1933:40, figs. 124—126 (pl. 12).
Phrurotimpus borealis: Kaston, 1938*c*:194; 1945:6, figs. 40, 41.

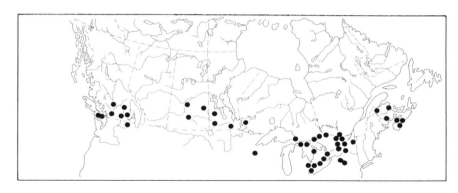

Map 39. Collection localities of *Phrurotimpus borealis*.

Male.   Total length approximately 2.55 mm; carapace 1.16 ± 0.07 mm long, 0.91 ± 0.07 mm wide (20 specimens measured). Carapace orange or orange yellow, with black lateral margins, black branching lines that radiate from dorsal groove area, and with pale band between posterior row of eyes and dorsal groove; sparsely covered with short iridescent setae. Chelicerae yellow, marked with black. Legs orange, marked with black; tibia I swollen, with ventral fringe of long dark setae and with six pairs of long pale ventral macrosetae (Fig. 248). Abdomen gray to black, with few indistinct pale chevrons posteriorly, with scutum covering most of dorsum, with epigastric scutum, and with scalelike recumbent iridescent setae. Femur of palpus with smooth hairy prominence (Fig. 243). Tibia of palpus slightly longer than wide, with small prominence on ventral side and with long straight retrolateral apophysis arising at tip of tibia; retrolateral apophysis bent at tip (Fig. 243). Tegulum convex, with excavation distally and with small spurlike apophysis; embolus short, twisted distally, with two small teeth at tip, arising broadly at tip of tegulum (Fig. 243).

Female.   Total length approximately 3.50 mm; carapace 1.34 ± 0.10 mm long, 1.08 ± 0.08 mm wide (20 specimens measured). General structure (Fig. 244) and color essentially as in male but tibia I (Fig. 245) sometimes with seven pairs of ventral macrosetae and abdomen paler and lacking scuta. Epigynum with convex hairy plate that is convex at posterior margin; copulatory openings large, deep, well-separated, situated at level of spermathecae (Fig. 247). Copulatory tubes short, broad; spermathecae in two parts, with anterior part long, saclike, much larger than posterior part, and with tubes connecting parts of spermathecae slightly curved (Fig. 246).

Comments.   Males of *P. borealis* are distinguished from those of the other species by having a straight retrolateral apophysis that arises distally on the palpal tibia and is expanded near its tip. Females of *borealis* differ from those of *P. alarius* by the more posterior position of the copulatory openings, and from those of *P. certus* by the elongate saclike anterior parts of the spermathecae. They differ from females of *P. dulcineus* by the slighter curvature of the tubes connecting the two parts of the spermathecae and by the larger copulatory openings.

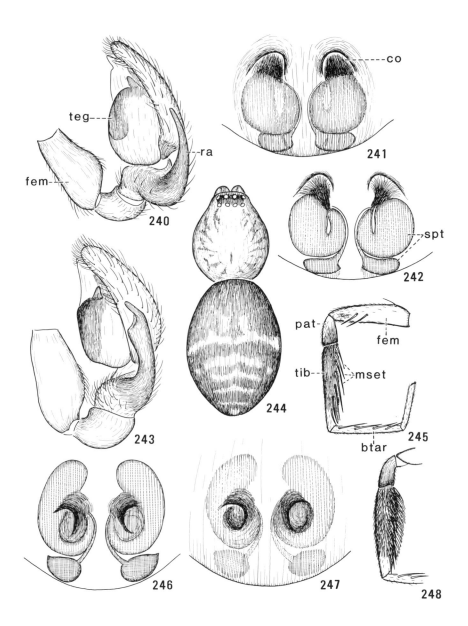

Figs. 240−248. Structures of *Phrurotimpus* spp. 240−242, *P. alarius*. 240, Palpus of male, retrolateral view; 241, Epigynum; 242, Spermathecae, ventral view. 243−248, *P. borealis*. 243, Palpus of male, retrolateral view; 244, Body of female, dorsal view; 245, Right leg I of female, prolateral view; 246, Spermathecae, ventral view; 247, Epigynum; 248, Right tibia I of male, prolateral view. *btar*, basitarsus; *co*, copulatory opening; *fem*, femur; *mset*, macrosetae; *pat*, patella; *ra*, retrolateral apophysis; *spt*, spermatheca; *teg*, tegulum; *tib*, tibia.

Range.   British Columbia to Nova Scotia, southward to Utah, northern Mexico, and North Carolina.

Biology.   Collections of *P. borealis* were made mainly by pitfall traps in the leaf litter of spruce, pine, oak, and beech−maple forests, in prairies, bogs, swamps, and meadows, on rocky hillsides, and under stones and beach debris. A few specimens have been taken inside houses. Mature males were collected from May to August, and mature females from May to November. Kaston (1948) mentioned a pair mating in early July.

Phrurotimpus dulcineus Gertsch

Figs. 249−253; Map 40

Phrurotimpus dulcineus Gertsch, 1941a:16, figs. 42−44.
Phrurotimpus minutus: Kaston, 1948:389, fig. 1389 (pl. 73) (female only).

Male.   Total length approximately 1.75 mm; carapace 0.79−0.84 mm long, 0.60−0.64 mm wide (five specimens measured). Carapace dark orange brown, with dark margin and with network of darker lines or bands. Chelicerae dark orange brown, marked with black. Legs orange or orange brown except distal end of femur, patella, and basal two-thirds of tibia of leg I, which are nearly black; tibia I somewhat swollen, without fringe of dark setae on ventral side. Abdomen gray, with small pale spot at posterior end, and with shiny orange gray scutum covering most of dorsum. Femur of palpus with small smooth hairy prominence on ventral side (Fig. 250). Tibia of palpus slightly longer than wide, with small ventral prominence and with single long stout retrolateral apophysis that arises at middle of tibia; tip of retrolateral apophysis slender, curved ventrad (Fig. 250). Tegulum convex, with small spurlike apophysis; embolus short, straight, slender, minutely bifid at tip (Fig. 250).

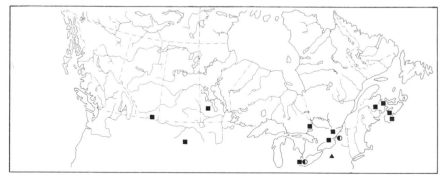

Map 40. Collection localities of *Phrurotimpus dulcineus* (◐), *P. certus* (■), and *P. minutus* (▲).

Female. Total length approximately 2.30 mm; carapace 0.93 ± 0.04 mm long, 0.72 ± 0.04 mm wide (10 specimens measured). General structure and color essentially as in male but tibia of leg I somewhat darker than other segments; abdomen with few pale chevrons, without scutum. Epigynum with rounded convex plate; copulatory openings conspicuous, situated at midlength of plate (Figs. 252, 253). Copulatory tubes short, funnellike; spermathecae each in two parts, with anterior part large and saclike and with posterior part smaller, ovoid, situated at genital groove, with both parts connected by strongly curved tube (Figs. 249, 251).

Comments. Males of *P. dulcineus* differ from those of other species by having a slender sinuous retrolateral apophysis that arises at the middle of the palpal tibia. Females of *dulcineus* differ from those of *P. alarius* by the more posterior position of the copulatory openings, and from females of *P. certus* by the long saclike anterior parts of the spermathecae. Females differ from those of *P. borealis* by the strongly curved tubes that connect the two parts of the spermathecae and by the smaller copulatory openings.

Range. Florida northward to Nebraska and southern Ontario.

Biology. Specimens of *P. dulcineus* have been collected by pitfall traps in tall marsh grass. Mature males were collected in January (Florida) and June, and mature females in March, July, and September.

Phrurotimpus certus Gertsch

Figs. 254–256; Map 40

Phrurotimpus certus Gertsch, 1941:17, figs. 47, 48.

Male. Total length approximately 2.05 mm; carapace 0.91 ± 0.03 mm long, 0.71 ± 0.04 mm wide (13 specimens measured). Carapace orange with black lateral margins and with several broken black lines radiating from dorsal groove area; with one pair of these black lines extending far anterolaterad; with recumbent iridescent setae. Chelicerae orange yellow, marked with black. Legs yellow orange, with patella and tibia I gray black with purple iridescence; tibia I with ventral fringe of black setae and with six pairs of long pale macrosetae. Abdomen gray to black with several paler chevrons, with dorsal scutum covering most of dorsum, with epigastric scutum, and with recumbent iridescent setae. Femur of palpus with smooth hairy prominence on ventral surface (Fig. 254). Tibia of palpus slightly longer than wide, with small ventral prominence and with single long straight retrolateral apophysis that arises from the base of the tibia and terminates in short stout part (Fig. 254). Tegulum convex, with distal excavation and with short spurlike apophysis; embolus short, pointed, without teeth, arising broadly at tip of tegulum (Fig. 254).

Female. Total length approximately 2.55 mm; carapace 0.99 ± 0.03 mm long, 0.78 ± 0.03 mm wide (17 specimens measured). General structure and color essentially as in male but abdomen lacking scuta. Epigynum with convex hairy

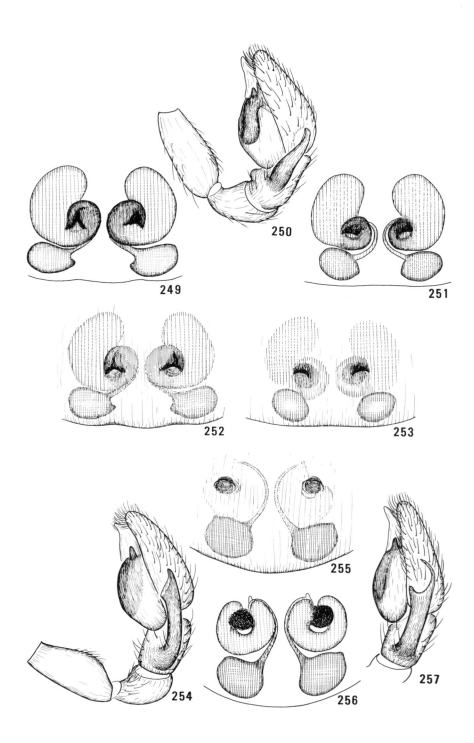

249

250

251

252

253

255

254

256

257

138

plate having convex posterior margin; copulatory openings large, well-separated, situated at level of spermathecae (Fig. 255). Copulatory tubes short, broad; spermathecae each in two parts, with anterior part rounded, slightly larger than posterior part (Fig. 256).

Comments. Males of *P. certus* are distinguished from those of *P. alarius* and *P. dulcineus* by the straight retrolateral apophysis that arises from the base of the palpal tibia and by the presence of a ventral fringe of dark setae on tibia I. They differ from males of *P. borealis* by having the retrolateral apophysis arise near the base of the tibia rather than at the distal end. They differ from males of *P. minutus* by having the tip of the retrolateral apophysis short and stout. Females of *certus* differ from those of *alarius* by the more posterior position of the copulatory openings and from females of *borealis* and *dulcineus* by the rounded anterior part and relatively large posterior part of the spermathecae.

Range. Alberta to Nova Scotia, southward to Wyoming, Alabama, and Virginia.

Biology. Collections of *P. certus* are mainly from pitfall traps in the leaf litter of oak or beech−maple forests and in meadows, prairies, and shrubby pastures. One specimen was found in a house. Mature males and females have been collected from April to November.

Phrurotimpus minutus (Banks)

Fig. 257; Map 40

Phrurolithus minutus Banks, 1892:22, figs. 67−67*b* (pl. 1); Emerton 1911:404, figs. 2−2*b* (pl. 6).
Phrurotimpus minutus: Kaston, 1948:389, fig. 1387 (pl. 73) (male only); Dondale & Redner 1979:266.

Male. Total length approximately 1.95 mm; carapace 0.87−0.93 mm long, 0.69−0.73 mm wide (four specimens measured). Carapace dark orange brown, with numerous darker branching lines, with dark margins, and with pale band along midline and pale area at dorsal groove. Chelicerae orange yellow. Legs orange yellow, with femur, patella, and basal four-fifths of tibia I dark brown to nearly black; tibia I swollen, without apparent fringe of black setae, with five or six pairs of long pale ventral macrosetae. Abdomen gray to black, with four pale chevrons on posterior two-thirds, and with shiny scutum covering most of dorsum. Palpal femur with smooth hairy ventral prominence. Palpal tibia slightly longer than wide, with small prominence on ventral side and with long straight slender

---

Figs. 249−257. Genitalia of *Phrurotimpus* spp. 249−253, *P. dulcineus*. 249, 251, Spermathecae, ventral view; 250, Palpus of male, retrolateral view; 252, 253, Epigynums. 254−256, *P. certus*. 254, Palpus of male, retrolateral view; 255, Epigynum; 256, Spermathecae, ventral view. 257, *P. minutus*. Palpus of male, retrolateral view.

retrolateral apophysis that arises near middle of tibia and terminates in long slender part (Fig. 257). Tegulum convex, with distal excavation and with short spurlike apophysis; embolus short, pointed, without teeth, arising rather broadly at tip of tegulum (Fig. 257).

Female. Total length approximately 2.20 mm; carapace 1.00 mm long, 0.84 mm wide (known only from single damaged female).

Comments. Males of *P. minutus* are distinguished from those of other species by the long straight slender retrolateral apophysis, which arises near the middle of the tibia and terminates in a long slender part. Females are not adequately diagnosed.

Range. Illinois to Massachusetts, southward to New Jersey.

Biology. Banks' (1892) original specimens were collected in September along two of the streams that drain southward into Cayuga Lake, New York. Another male was collected under straw in a field in Massachusetts in April, and the others were collected in May or June.

Genus *Scotinella* Banks

Spiders of the genus *Scotinella* are small, somewhat antlike, and secretive. They are poorly known biologically, though they may associate in some way with ants and thereby obtain a degree of immunity from predation. They may be seen running with ants during daylight, and some, such as *S. britcheri* (Petrunkevitch), live within the ant nest itself. Representatives of a few species such as *S. brittoni* (Gertsch), which are shiny and black, quite closely resemble the ants in whose nest they are found, in this instance *Crematogaster lineolata* (Say). The spiders' thin legs, antlike gait, and method of holding the front legs forward in the position of antennae further reinforce this mimicry. The mechanism by which the spiders are permitted to dwell without apparent harm in these nests is unknown.

Description. Total length 1.75–2.65 mm. Carapace (Fig. 12) pale gray, orange brown, or brown, often veined with black, usually without conspicuous setae except few short erect ones in eye area, approximately ovoid in dorsal view, usually highest at anterior end of dorsal goove, with surface minutely pitted; dorsal groove shallow, inconspicuous. Eyes usually prominent, round, uniform in size, arranged in two transverse rows; both rows essentially straight or with posterior row slightly procurved; posterior row as long as anterior row or slightly longer; eyes of both rows uniformly spaced or with median eyes slightly closer to each other than to laterals, occasionally with median eyes slightly closer to laterals than to each other; median eyes of either row sometimes angular or ovoid in outline. Chelicerae short, stout, tapered, orange brown, often with black markings; promargin of fang furrow with two or three minute teeth; retromargin with two or three minute teeth. Palp-coxal lobes approximately as wide as long, convex along lateral margin. Legs yellow orange to dark brown, rather slender (Fig. 12), with thin claw tufts and scopulae; femur I without dorsal macrosetae, with two prolaterals (rarely one or three); tibia I with five or six pairs of long overlapping ventral macrosetae; basitarsus I with four pairs of long overlapping ventral macrosetae (Fig. 12); trochanter IV without distal notch on ventral side. Abdomen ovoid in dorsal view; dark orange, gray, or black, sometimes with pattern of one or more pale chevrons; with short inconspicuous setae, but without cluster of long erect setae at anterior end; abdomen of male (and sometimes female) with large shiny dorsal scutum and usually with epigastric scutum. Femur of male palpus with hooked prominence on ventral side (e.g., Figs. 265, 296); tibia slightly longer than wide, wih small ventral prominence and with strong retrolateral apophysis composed of two processes (e.g., Figs. 262, 265, 277); tegulum rounded, convex, without apophysis; embolus short, usually expanded at base and slender distad, arising at distal end of tegulum (e.g., Figs. 262, 275, 285). Epigynum of female usually with plate elongate, rarely rounded, convex, often with conspicuous paired atrial depressions in anterior half (e.g., Figs. 263, 266, 290); copulatory openings usually small, inconspicuous, often slitlike, located in margins of atrial depressions or appearing as simple cavities in epigynal plate (e.g., Figs. 263, 266, 290). Copulatory tubes short to long, slender, often arched laterad; spermathecae ovoid or round in outline, located posterior or posterolateral to copulatory openings, with slender club-shaped spermathecal organ at anterior end (e.g., Figs. 264, 267, 280).

Comments. Specimens of *Scotinella* spp. are distinguished from those of *Cheiracanthium* spp., *Clubiona* spp., and *Clubionoides* spp. by the absence of a constriction on the lateral margins of the palp-coxal lobes. They differ from specimens of *Castianeira* spp., *Agroeca* spp., and *Trachelas* spp. by having the anterior row of eyes straight or nearly so and by having four pairs of ventral macrosetae on basitarsus I. They differ from specimens of *Phrurotimpus* spp. by lacking dorsal macrosetae on femur I, and by having a hooked ventral prominence on the male palpal femur, a retrolateral apophysis having dorsal and ventral processes on the male palpal tibia, and less conspicuous copulatory openings and simpler spermathecae in females.

Gertsch (1979) and Kaston (1972, 1978) indicated that the group of species represented by *S. pugnata* is sufficiently distinct from the Old World genus *Phrurolithus*, in which most of them were originally described, to comprise a separate genus. The oldest available name for such a group is *Scotinella*. We follow Gertsch and Kaston in its use even though many of the specific names have not been actually published in that combination. Current knowledge of the species is meagre, but revision may reveal 35 or more described species in *Scotinella*. Eleven species are represented or assumed to be represnted in Canada.

Key to species of *Scotinella*

(Female of *S. deleta* is unknown)

1. Male 2
   Female 12
2(1). Dorsal process of retrolateral tibial apophysis extending much farther distad than ventral process; retrolateral apophysis more than one and one-half times as long as wide in retrolateral view (e.g., Figs. 265, 274, 288) 3

Dorsal process of retrolateral tibial apophysis extending only slightly farther distad than ventral process; retrolateral apophysis less than one and one-half times as long as wide in retrolateral view (Figs. 296, 300, 302, 306) . . . 9

3(2). Dorsal and ventral processes of retrolateral tibial apophysis slender, pointed, not excavated or toothed (Figs. 265, 270, 274). Embolus broad and angular at base (Figs. 262, 268, 271) 4

Dorsal and/or ventral processes of retrolateral tibial apophysis thickened, blunt, excavated, or toothed (e.g., Figs. 275, 284, 288, 292). Embolus more slender (e.g., Figs. 275, 281, 285) 6

4(3). Distal part of embolus straight (retrolateral view, Fig. 270) . . . .
. . . . *divesta* (**Gertsch**) (p. 145)

Distal part of embolus curved (Figs. 265, 274) 5

5(4). Embolus gradually narrowed (ventral view, Fig. 262) . . . .
. . . . *pugnata* (**Emerton**) (p. 147)

Embolus abruptly narrowed at midlength (ventral view, Fig. 271) . . . .
. . . .*sculleni* (**Gertsch**) (p. 149)

6(3). Ventral process of retrolateral tibial apophysis short, with one or two minute teeth (Fig. 275) . . . .*redempta* (**Gertsch**) (p. 150)

Ventral process of retrolateral tibial apophysis longer, not toothed (Figs. 281, 285, 289) 7

7(6). Embolus broad at base, tapered and curved distad (Figs. 285, 289). Tip of dorsal process of retrolateral tibial apophysis curved dorsad (Figs. 288, 292) . . 8

Embolus slender throughout its length (Fig. 281). Tip of dorsal process of retrolateral tibial apophysis curved ventrad (Fig. 284) . . . .
. . . . *madisonia* **Levi** (p. 151)

8(7). Dorsal process of retrolateral tibial apophysis long, without excavation near tip (Fig. 288) . . . . *fratrella* (**Gertsch**) (p. 153)

Dorsal process of retrolateral tibial apophysis shorter, with excavation near tip (Fig. 292) . . . . *britcheri* (**Petrunkevitch**) (p. 154)

9(2). Dorsal process of retrolateral tibial apophysis approximately same thickness throughout, tapered only near tip (Fig. 300) . . . .
. . . . *minnetonka* (**Chamberlin & Gertsch**) (p. 156)

Dorsal process of retrolateral tibial apophysis thickest at base, gradually tapered to tip (Figs. 296, 302, 306) 10

10(9). Dorsal process of retrolateral tibial apophysis blunt at tip (Fig. 296) . . . .
. . . . *brittoni* (**Gertsch**) (p. 159)

Dorsal process of retrolateral tibial apophysis finely pointed at tip (Figs. 302, 306) 11

11(10). Embolus with slender distal part longer than basal part (Fig. 301); tip of dorsal process of retrolateral tibial apophysis extending farther distad than ventral process (Fig. 302) . . . . *deleta* (**Gertsch**) (p. 160)

Embolus with slender distal part approximately as long as thicker basal part (Fig. 303); tip of dorsal process of retrolateral tibial apophysis extending as far distad as ventral process (Fig. 306) . . . . *divinula* (**Gertsch**) (p. 161)

12(1). Copulatory openings placed one anterior to the other (Figs. 286, 290) 13

Copulatory openings both placed at same level (e.g., Figs. 263, 266, 282) . 14

13(12). Copulatory tubes with bulbous swellings near openings (Fig. 287) . . . .
. . . . *fratrella* (**Gertsch**) (p. 153)

Copulatory tubes without swellings near openings (Fig. 291) . . . .
. . . . *britcheri* (**Petrunkevitch**) (p. 154)

14(12). Epigynum with paired dark smooth swellings at copulatory openings (Figs. 263, 269) 15

Epigynum without paired dark smooth swellings 16

Clé des espèces de *Scotinella*

(Femelle de *S. deleta* inconnue)

1. Mâle 2
Femelle 12
2(1). Processus dorsal de l'apophyse tibiale rétrolatérale se prolongeant beaucoup plus loin distalement que le processus ventral; apophyse rétrolatérale plus qu'une fois et demie plus longue que large en vue rétrolatérale (p. ex., fig. 265, 274 et 288) 3
Processus dorsal de l'apophyse tibiale rétrolatérale se prolongeant à peine plus loin distalement que le processus ventral; apophyse rétrolatérale moins de une fois et demie plus longue que large en vue rétrolatérale (fig. 296, 300, 302 et 306) 9
3(2). Processus dorsal et ventral de l'apophyse tibiale rétrolatérale étroits, pointus, non excavés ou dentés (fig. 265, 270 et 274). Embolus large et angulaire à la base (fig. 262, 268 et 271) 4
Processus dorsal et (ou) ventral de l'apophyse tibiale rétrolatérale épaissis, obtus, excavés ou dentés (p. ex., fig. 275, 284, 288 et 292). Embolus plus étroit (p. ex., fig. 275, 281 et 285) 6
4(3). Partie distale de l'embolus droite (vue rétrolatérale, fig. 270) . . . .
. . . . *divesta* (**Gertsch**) (p. 145)
Partie distale de l'embolus courbée (fig. 265 et 274) 5
5(4). Embolus graduellement rétréci (vue ventrale, fig. 262) . . . .
. . . . *pugnata* (**Emerton**) (p. 147)
Embolus brusquement rétréci en son milieu (vue ventrale, fig. 271) . . . .
. . . . *sculleni* (**Gertsch**) (p. 149)
6(3). Processus ventral de l'apophyse tibiale rétrolatérale court, avec une ou deux dents minuscules (fig. 275) . . . . *redempta* (**Gertsch**) (p. 150)

Scotinella divesta (Gertsch)

Figs. 266−268, 270; Map 41

Phrurolithus divestus Gertsch, 1941a:6, figs. 19−21; Kaston 1948:393, figs. 1399, 1400 (pl. 74).

Male. Total length approximately 2.05 mm; carapace 0.89−0.96 mm long, 0.74−0.81 mm wide (five specimens measured). Carapace orange brown, darker in eye area, with indistinct darker lines radiating from dorsal groove area. Chelicerae dark orange. Legs dark orange. Abdomen dark orange brown suffused with black; with scutum covering entire dorsum. Femur of palpus with hooked apophysis on ventral surface. Tibia of palpus with retrolateral apophysis nearly twice as long as wide; dorsal process of apophysis extending much farther distad than ventral process, and neither process excavated or toothed (Fig. 270). Tegulum convex, without apophysis; embolus short, with broad angular base and tapered distal part (Fig. 268).

Female. Total length approximately 2.35 mm; carapace 0.92 ± 0.02 mm long, 0.77 ± 0.02 mm wide (17 specimens measured). General structure and color essentially as in male but abdomen lacking dorsal scutum. Epigynum with

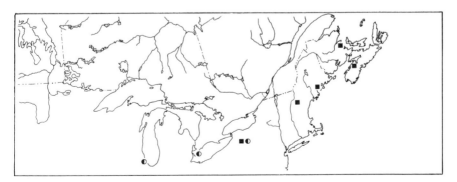

Map 41. Collection localities of *Scotinella divesta* (■) and *S. redempta* (◐).

145

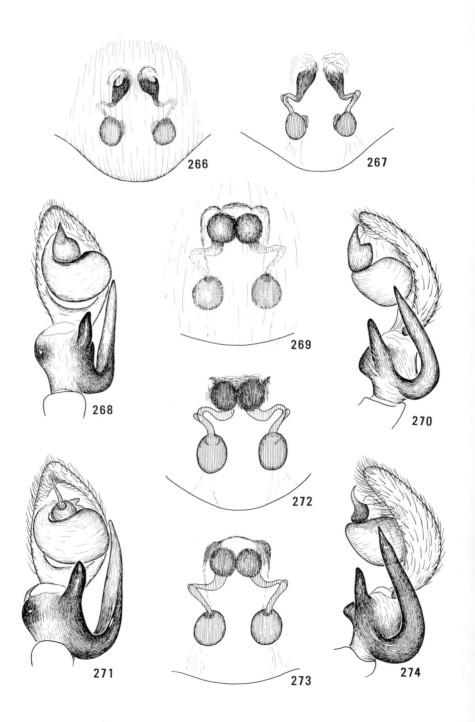

elongate plate convex along posterior margin; copulatory openings small, located at margins of small depressions (Fig. 266). Copulatory tubes short, slender, abruptly arched laterad; spermathecae small, ovoid, with slender club-shaped spermathecal organs (Fig. 267).

Comments. Specimens of *S. divesta* are distinguished from those of the other species by the following combination of characters: dorsal process of retrolateral tibial apophysis in males extending much farther distad than ventral process and both dorsal and ventral processes slender, tapered, and without teeth or excavations; distal part of male embolus straight rather than curved; epigynum without paired smooth swellings near copulatory openings, which are placed at same level on epigynal plate; copulatory tubes abruptly arched laterad.

Range. Northern New York to Nova Scotia, southward to Massachusetts.

Biology. Collections of *S. divesta* have been taken by pitfall traps in sphagnum bogs and in spruce−fir forests, and by hand under stones and debris in fields and beaches and under logs on the ground. Mature males and females have been found from March to November.

Scotinella pugnata (Emerton)

Figs. 12, 258−265; Map 42

Phrurolithus pugnatus Emerton, 1890:188, figs. 6−6c (pl. 6); Gertsch 1941a:4, figs. 22−24; Kaston 1948:390, figs. 1357−1359 (pl. 71), fig. 1390 (pl. 73).

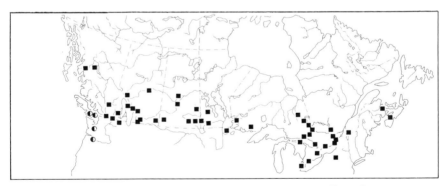

Map 42. Collection localities of *Scotinella pugnata* (■) and *S. sculleni* (◐).

Figs. 266−274. Genitalia of *Scotinella* spp. 266−268, 270, *S. divesta*. 266, Epigynum; 267, Spermathecae, ventral view; 268, Palpus of male, ventral view; 270, Palpus of male, retrolateral view. 269, 271−274, *S. sculleni*. 269, Epigynum; 271, Palpus of male, ventral view; 272, 273, Spermathecae, ventral view; 274, Palpus of male, retrolateral view.

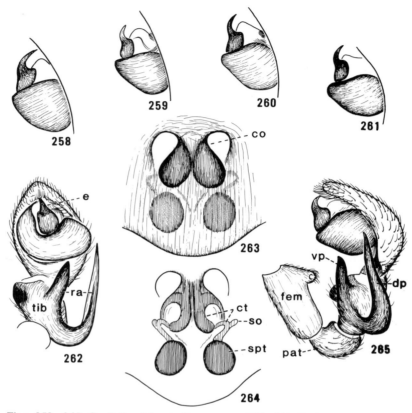

Figs. 258–265. Genitalia of *Scotinella pugnata*. 258–261, Emboli of males, retrolateral view; 262, Palpus of male, ventral view; 263, Epigynum; 264, Spermathecae, ventral view; 265, Palpus of male, retrolateral view. *co*, copulatory opening; *ct*, copulatory tube; *dp*, dorsal process; *e*, embolus; *fem*, femur; *pat*, patella; *ra*, retrolateral apophysis; *so*, spermathecal organ; *spt*, spermatheca; *tib*, tibia; *vp*, ventral process.

Male. Total length approximately 2.20 mm; carapace 0.98 ± 0.05 mm long, 0.83 ± 0.04 mm wide (20 specimens measured). Carapace orange brown, finely veined with black. Chelicerae orange brown, marked with black. Legs yellow orange, with indistinct black longitudinal bands along prolateral and retrolateral surfaces of femora and patellae. Abdomen gray to black, with one or more indistinct pale chevrons, with shiny scutum covering most of dorsum and usually with epigastric scutum. Femur of palpus with hooked prominence on ventral side. Tibia of palpus slightly longer than wide, with small ventral prominence and with retrolateral apophysis approximately twice as long as wide; dorsal process of retrolateral apophysis extending much farther distad than ventral process; dorsal and ventral processes slender, tapered, not toothed or excavated (Fig. 265). Tegulum convex, without apophysis; embolus with broad angular base and gradually narrowed distal part (Figs. 258–262).

Female. Total length approximately 2.45 mm; carapace 0.98 ± 0.08 mm long, 0.82 ± 0.06 mm wide (20 specimens measured). General structure and color essentially as in male but abdomen without scuta. Epigynum with elongate plate convex along posterior margin; copulatory openings inconspicuous, leading to large paired dark smooth swellings (Fig. 263). Copulatory tubes rather long, curving together at midline then abruptly arching laterad; spermathecae rather large, ovoid, with slender spermathecal organs (Fig. 264).

Comments. Specimens of *S. pugnata* are distinguished from those of other species by the following combination of characters: dorsal process of male retrolateral tibial apophysis extending farther distad than ventral process, slender, tapered, without teeth or excavations; distal part of embolus curved, gradually narrowed; both copulatory openings of female placed at same level on epigynal plate, leading to large paired smooth swellings. The embolus and epigynal swellings vary in degree of expression, appearing to approach the condition found in specimens of *S. sculleni* in the west. Further study is needed.

Range. British Columbia to Nova Scotia, southward to Utah and Delaware.

Biology. Most collections of *S. pugnata* have been made by pitfall traps placed in plant litter in fields, marshes, calcareous and sphagnum bogs, hedges, and prairies or in oak, aspen, and spruce – fir forests. A few were taken under stones along the shores of lakes or in sweep nets from pine foliage. Adults of both sexes have been collected from March to October.

Scotinella sculleni (Gertsch)

Figs. 269, 271 – 274; Map 42

Phrurolithus sculleni Gertsch, 1941a:8, figs. 25 – 27.

Male. Total length approximately 2.10 mm; carapace 0.96 mm long, 0.82 mm wide (one specimen measured). Carapace orange brown, veined with black. Chelicerae yellow brown, marked with black. Legs orange brown or yellow brown. Abdomen dark orange suffused with black, with pattern of pale chevrons faintly indicated; with shiny scutum covering most of dorsum, and with pale epigastric scutum. Femur of palpus with hooked apophysis on ventral side. Tibia of palpus slightly longer than wide, with small ventral prominence and with large retrolateral apophysis approximately twice as long as wide; dorsal process of retrolateral apophysis extending much farther distad than ventral process; dorsal and ventral processes slender, tapered, without teeth or excavations (Fig. 274). Tegulum convex, without apophysis; embolus broad and angular at base, abruptly narrowed at midlength (Fig. 271).

Female. Total length approximately 2.40 mm; carapace 1.02 mm long, 0.87 mm wide (one specimen measured). General structure and color essentially as in male but carapace lacking dark markings and abdomen lacking dorsal scutum; chevron pattern rather distinct. Epigynum with elongate plate convex along posterior margin; copulatory openings round or slitlike, leading to small paired dark smooth swellings (Fig. 269). Copulatory tubes rather long, slender, abruptly arched laterad; spermathecae small, ovoid, with slender spermathecal organs (Figs. 272, 273).

Comments. Specimens of *S. sculleni* are distinguished from those of other species by the following combination of characters: dorsal process of male retrolateral tibial apophysis extending farther distad than ventral process, slender, tapered, without teeth or excavations; distal part of male embolus curved, abruptly narrowed; both copulatory openings of female placed at same level on epigynal plate, leading to small paired dark swellings. Specimens of *sculleni* may be difficult to distinguish from western specimens of *S. pugnata* owing to variability in the genital structure of the latter.

Range. Coastal British Columbia, Washington, and Oregon.

Biology. A female was collected in a pitfall trap in a grassy clearing within a maple stand.

Scotinella redempta (Gertsch)

Figs. 275–280; Map 41

Phrurolithus redemptus Gertsch, 1941a:2, figs. 15, 16, 18.
Scotinella redempta: Roddy, 1957:288.

Male. Total length approximately 2.45 mm; carapace 1.12 ± 0.06 mm long, 0.93 ± 0.05 mm wide (14 specimens measured). Carapace orange brown with indistinct darker lines radiating from dorsal groove area. Chelicerae orange brown. Legs dark orange. Abdomen dark orange suffused with black, with pattern of pale chevrons sometimes faintly indicated; with shiny scutum covering most of dorsum. Femur of palpus with hooked apophysis on ventral surface. Tibia of palpus slightly longer than wide, with small ventral prominence and with large retrolateral apophysis nearly twice as long as wide; dorsal process of retrolateral apophysis extending much farther distad than ventral process, broad at base, smoothly tapered distad; ventral process of retrolateral apophysis short, blunt, bearing one or two minute teeth (Figs. 275, 277). Tegulum convex, without apophysis; embolus short, slender (Fig. 275).

Female. Total length approximately 2.65 mm; carapace 1.12 ± 0.06 mm long, 0.92 ± 0.05 mm wide (20 specimens measured). General structure and color essentially as in male but abdomen lacking dorsal scutum and with chevron pattern distinct; tibia I with six to eight pairs of ventral macrosetae. Epigynum with elongate plate convex along posterior margin; copulatory openings small, close together, nearly round, located in anterior part of epigynal plate (Figs. 276, 279). Copulatory tubes slender, extending first anterior to copulatory openings then posterior, not sinuous or abruptly arched laterad; spermathecae elliptical, oblique (Figs. 278, 280).

150

Comments. Specimens of *S. redempta* are distinguished from those of other species by the following combination of characters: dorsal process of male retrolateral tibial apophysis extending much farther distad than ventral process, and ventral process short, blunt, bearing one or two minute teeth; embolus of male slender, with curved distal part; both copulatory openings of female placed at same level on epigynal plate, not leading to paired dark swellings; copulatory tubes straight in posterior part, extending anterior to copulatory openings then posterior.

Range. Kansas, Alabama, and North Carolina, northward to southern Ontario.

Biology. Specimens of *S. redempta* have been collected by litter sifters in deciduous forest. Mature males were collected in May, and mature females in May and July.

Scotinella madisonia Levi

Figs. 281−284; Map 43

Scotinella madisonia Levi, 1951:27, figs. 23, 27.

Male. Total length approximately 2.25 mm; carapace 0.99−1.16 mm long, 0.86−0.94 mm wide (nine specimens measured). Carapace dark orange brown, finely veined with black. Chelicerae yellow brown, marked with black. Legs brown orange, with diffuse black pigment along prolateral surfaces of leg femora and along ventral surfaces of tibiae and basitarsi III and IV. Abdomen nearly black, with thick transverse pale band near middle; with shiny scutum covering dorsum, and with epigastric scutum. Femur of palpus with hooked apophysis on ventral surface. Tibia of palpus approximately as broad as long, with large retrolateral apophysis approximately twice as long as wide; dorsal process of retrolateral apophysis thickened, extending much farther distad than ventral process, angled ventrad at tip (Fig. 284). Tegulum convex, without apophysis; embolus slender, tapered (Fig. 281).

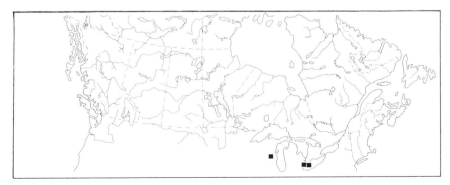

Map 43. Collection localities of *Scotinella madisonia*.

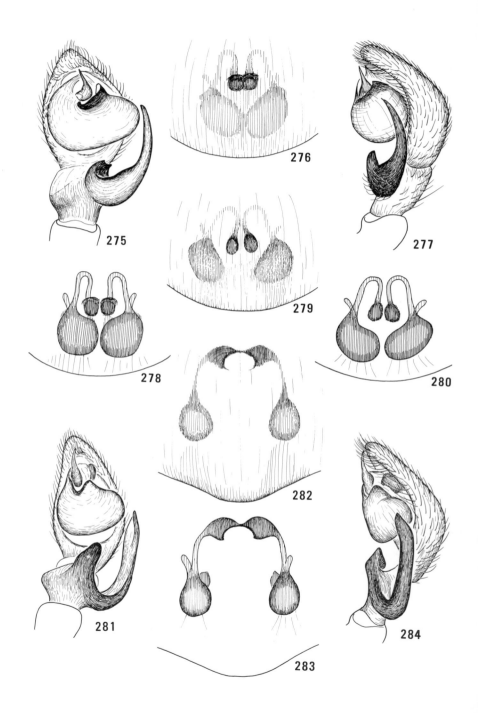

275

276

277

278

279

280

281

282

283

284

Female. Total length approximately 2.35 mm; carapace 0.99 ± 0.06 mm long, 0.85 ± 0.04 mm wide (20 specimens measured). General structure and color essentially as in male but abdomen darker and lacking scuta. Epigynum with elongate plate convex along posterior margin; copulatory openings inconspicuous, located in margins of shallow anterior depressions (Fig. 282). Copulatory tubes slender, nearly straight; spermathecae ovoid, small, placed well anterior to genital groove, with slender club-shaped spermathecal organs (Fig. 283).

Comments. Specimens of *S. madisonia* are distinguished from those of other species in the genus by the following combination of characters: dorsal process of retrolateral tibial apophysis in males extending much farther distad than ventral process, thickened, angled ventrad at tip; embolus of male slender, slightly curved; copulatory openings of females both at same level on epigynal plate, not leading to paired dark swellings; posterior parts of copulatory tubes nearly straight, not extending anteriad before extending posteriad; spermathecae small, placed well anterior to genital groove.

Range. Wisconsin and southern Ontario.

Biology. Specimens of *S. madisonia* have been collected by pitfall traps in the leaf litter of a relict prairie in which pin oaks were growing. Others were collected under sticks and other ground debris, and one was found in a house. Mature males and females were found from April to August.

Scotinella fratrella (Gertsch)

Figs. 285−288; Map 44

Phrurolithus fratrellus Gertsch, 1935:6, fig. 21; Barrows & Ivie 1942:20, figs. 6, 7; Penniman 1978:127, figs. 1−4, 9−12, 15−17.

Male. Total length approximately 1.75 mm; carapace 0.80 ± 0.02 mm long, 0.66 ± 0.02 mm wide (20 specimens measured)(from Penniman 1978). Carapace yellow orange, darker within eye area. Eyes rather large. Chelicerae orange. Legs yellow orange. Abdomen gray, with three or four indistinct paler chevrons; with indistinct scutum covering most of dorsum. Femur of palpus with hooked apophysis on ventral side. Tibia of palpus slightly longer than wide, with small ventral prominence and large retrolateral apophysis approximately twice as long as wide; dorsal process of retrolateral apophysis extending much farther distad than ventral process, tapered and curved dorsad near tip; ventral process with small excavation on dorsal side (Fig. 288). Tegulum convex, without apophysis; embolus curved, smoothly tapered from base to tip (Fig. 285).

Figs. 275−284. Genitalia of *Scotinella* spp. 275−280, *S. redempta*. 275, Palpus of male, ventral view; 276, 279, Epigynums; 277, Palpus of male, retrolateral view; 278, 280, Spermathecae, ventral view. 281−284, *S. madisonia*. 281, Palpus of male, ventral view; 282, Epigynum; 283, Spermathecae, ventral view; 284, Palpus of male, retrolateral view.

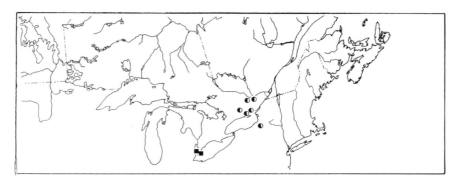

Map 44. Collection localities of *Scotinella fratrella* (■) and *S. britcheri* (◐).

Female. Total length approximately 2.00 mm; carapace 0.83 ± 0.02 mm long, 0.69 ± 0.02 mm wide (19 specimens measured) (from Penniman 1978). General structure and color essentially as in male but abdomen lacking dorsal scutum. Epigynum with rounded plate convex at posterior margin; copulatory openings placed at different levels on epigynal plate, small, inconspicuous (Fig. 286). Copulatory tubes with bulbous swellings near copulatory openings, otherwise slender, extending anterolaterad then posteriad; spermathecae kidney-shaped, not constricted at middle (Fig. 287).

Comments. Specimens of *S. fratrella* are distinguished from those of other species by the following combination of characters: size small (see Penniman 1978); dorsal process of male retrolateral tibial apophysis not excavated; copulatory openings of female placed one anterior to the other; copulatory tubes with bulbous swellings; spermathecae kidney-shaped, without constriction at middle.

Range. Texas to southern Ontario.

Biology. Specimens of *S. fratrella* have been collected by pitfall traps in fields and deciduous forests and under pin oaks in a relict prairie. Mature males were collected from April to July and in October, and mature females from April to September. Penniman (1978) notes a statistical relationship between high pitfall catches of *S. fratrella* and the waxing or waning of the moon.

Scotinella britcheri (Petrunkevitch)

Figs. 289–292; Map 44

Phrurolithus britcheri Petrunkevitch, 1910:217, fig. 23 (pl. 22); Kaston 1948:392, fig. 1396 (pl. 74); Penniman 1978:129, figs. 5–8, 13, l4.
  Scotinella britcheri: Gertsch, 1979:212.

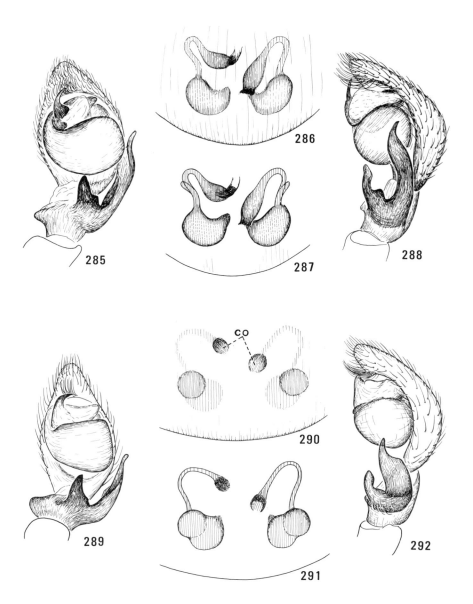

Figs. 285–292. Genitalia of *Scotinella* spp. 285–288, *S. fratrella*. 285, Palpus of male, ventral view; 286, Epigynum; 287, Spermathecae, ventral view; 288, Palpus of male, retrolateral view. 289–292, *S. britcheri*. 289, Palpus of male, ventral view; 290, Epigynum; 291, Spermathecae, ventral view; 292, Palpus of male, retrolateral view. *co*, copulatory openings.

Male.   Total length approximately 1.95 mm; carapace 0.94, 0.96 mm long, 0.80 mm wide (two specimens measured). Carapace pale orange brown or pale gray, sometimes with few darker bands radiating from dorsal groove area. Eyes rather small. Chelicerae pale orange brown or pale gray. Legs off-white or pale orange brown. Abdomen pale yellow brown, without chevrons, with inconspicuous scutum covering most of dorsum. Femur of palpus with hooked apophysis on ventral side. Tibia of palpus slightly longer than wide, with small ventral prominence and with retrolateral apophysis that is slightly more than one and one-half times as long as wide; dorsal process of retrolateral apophysis extending farther distad than ventral process, deeply excavated on dorsal margin near tip, with tip curved dorsad; ventral process of retrolateral apophysis rounded at tip, with small excavation near base (Fig. 292). Tegulum convex, without apophysis; embolus short, slender, curved, tapered from base to tip (Fig. 289).

Female.   Total length approximately 2.20 mm; carapace 0.98 ± 0.04 mm long, 0.80 ± 0.03 mm wide (20 specimens measured). General structure and color essentially as in male but abdomen lacking scutum. Epigynum with rounded plate convex along posterior margin; copulatory openings small but rather conspicuous, placed one anterior to the other (Fig. 290). Copulatory tubes slender, without swellings, extending anterolaterad then posterolaterad; spermathecae rounded, with constriction near middle (Fig. 291).

Comments.   Specimens of *S. britcheri* are distinguished from those of other species by the following combination of characters: size large (see Penniman 1978); dorsal process of retrolateral tibial apophysis in males relatively short, excavated; copulatory openings of female placed one anterior to the other; copulatory tubes without bulbous swellings; spermathecae rounded, with constriction near middle.

Range.   North Carolina to southern Ontario.

Biology.   Most specimens of *S. britcheri* were collected from the subterranean nests of the ant *Acanthomyops latipes* (Walsh). These ants nest under stones in fields and pastures, and their plump orange yellow bodies and slow locomotion contrast with the sleek pale bodies and swift running of the spiders. A few females of *britcheri* have been collected in pitfall traps. Mature males were taken in September, and mature females from May to September.

Scotinella minnetonka (Chamberlin & Gertsch)

Figs. 297–300; Map 45

Phruronellus minnetonka Chamberlin & Gertsch, 1930:139, figs. 13–15 (pl. 3).
Scotinella minnetonka: Levi, 1951:28.

Male.   Total length approximately 2.05 mm; carapace 0.87–1.01 mm long, 0.72–0.38 mm wide (seven specimens measured). Carapace orange brown, finely veined with black. Chelicerae brown yellow, marked with black. Legs

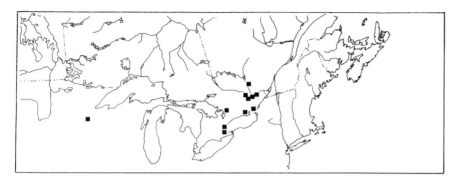

Map 45. Collection localities of *Scotinella minnetonka*.

yellow orange. Abdomen gray to black, with scutum covering entire dorsum and with epigastric scutum. Femur of palpus with hooked prominence on ventral side. Tibia of palpus slightly longer than wide, with small ventral prominence and with large retrolateral apophysis less than one and one-half times as long as wide; dorsal process of retrolateral apophysis extending slightly farther distad than ventral process, approximately same thickness throughout, tapered only near tip (Fig. 300). Tegulum convex, without apophysis; embolus with rounded base, tapered distad (Fig. 297).

Female. Total length approximately 2.25 mm; carapace 0.92 ± 0.04 mm long, 0.77 ± 0.03 mm wide (14 specimens measured). General structure and color essentially as in male but abdomen lacking scuta. Epigynum with elongate plate convex along posterior margin; copulatory openings inconspicuous, placed at posterior margins of large paired atrial depressions (Fig. 298). Copulatory tubes short, slender, abruptly arched laterad; spermathecae rather small, ovoid (Fig. 299).

Comments. Specimens of *S. minnetonka* are distinguished from those of other species by the following combination of characters: dorsal process of retrolateral apophysis on male palpal tibia extending only slightly farther distad than ventral process, tapered only at tip; retrolateral apophysis less than one and one-half times as long as wide; both copulatory openings of female placed at same level on epigynal plate, located at posterior margins of large atrial depressions; epigynum without paired dark smooth swellings.

Range. Minnesota to southern Ontario, southward to Mississippi.

Biology. Specimens of *S. minnetonka* have been collected in pitfall traps set in pastures, meadows, swamps, and oak–maple forests, and by hand under stones and in deciduous leaf litter. Mature males were collected from April to August, and mature females from June to September. One female was collected under snow in midwinter in Gatineau Park, Que.

157

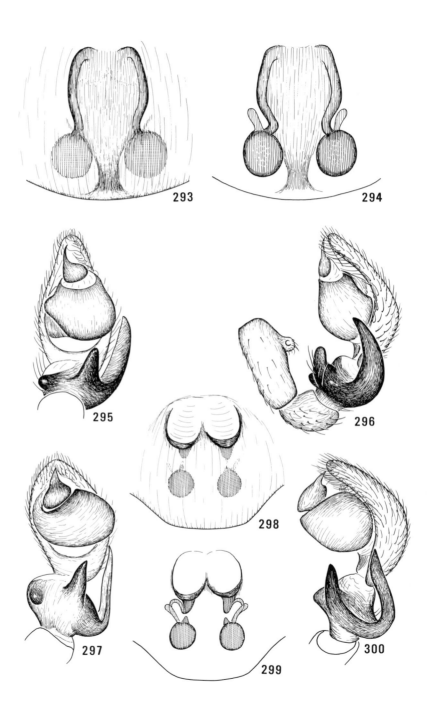

Scotinella brittoni (Gertsch)

Figs. 293–296; Map 46

Phrurolithus brittoni Gertsch, 1941a:14, figs. 34–36; Kaston 1948:391, figs. 1394, 1395 (pl. 74).

Male. Total length approximately 2.30 mm; carapace 1.03 ± 0.05 mm long, 0.89 ± 0.03 mm wide (18 specimens measured). Carapace dark brown, finely veined with black. Chelicerae brown yellow, marked with black. Legs dark brown, with indistinct longitudinal black bands along prolateral and retrolateral surfaces. Abdomen gray to black; with shiny scutum covering entire dorsum and with epigastric scutum. Femur of palpus with hooked apophysis on ventral side. Tibia of palpus with small ventral prominence and with large retrolateral apophysis less than one and one-half times as long as wide; dorsal process of retrolateral apophysis broad at base, gradually tapered to blunt tip (Fig. 296). Tegulum convex, without apophysis; embolus broad at base, with slender tapered distal part (Fig. 295).

Female. Total length approximately 2.50 mm; carapace 1.03 ± 0.04 mm long, 0.89 ± 0.03 mm wide (20 specimens measured). General structure and color essentially as in male but dorsal scutum less extensive and body usually dark orange brown. Epigynum with elongate plate convex along posterior margin; copulatory openings inconspicuous, placed at lateral margins of small anterior median depression (Fig. 293). Copulatory tubes long, slender, slightly curved,

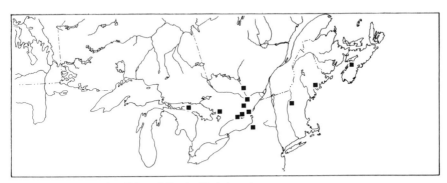

Map 46. Collection localities of *Scotinella brittoni*.

Figs. 293–300. Genitalia of *Scotinella* spp. 293–296, *S. brittoni*. 293, Epigynum; 294, Spermathecae, ventral view; 295, Palpus of male, ventral view; 296, Palpus of male, retrolateral view. 297–300, *S. minnetonka*. 297, Palpus of male, ventral view; 298, Epigynum; 299, Spermathecae, ventral view; 300, Palpus of male, retrolateral view.

extending first laterad then posteriad; spermathecae rather large, ovoid, placed short distance anterior to posterior margin of epigynum, with slender spermathecal organ (Fig. 294).

Comments.   Specimens of *S. brittoni* are distinguished from those of other species by the following combination of characters: dorsal process of retrolateral tibial apophysis in males extending slightly farther distad than ventral process, thickest at base, gradually tapered to blunt tip; retrolateral apophysis less than one and one-half times as long as wide; copulatory openings of females both at same level on epigynal plate; epigynum without paired dark swellings; copulatory tubes slightly curved, extending laterad before extending posteriad; spermathecae rather large, placed short distance anterior to posterior margin of epigynal plate; female with dorsal scutum on abdomen.

Range.   Southern Ontario to Nova Scotia, southward to Missouri and Maryland.

Biology.   Specimens of *S. brittoni* have been collected by pitfall traps in stony pastures and from nests of the ant *Crematogaster lineolata* (Say). Mature males were collected from July to October, and mature females from April to September.

Scotinella deleta (Gertsch)

Figs. 301, 302; Map 47

Phrurolithus deletus Gertsch, 1941a:10, figs. 28, 29.

Male.   Total length approximately 1.90 mm ; carapace 0.87 mm long, 0.76 mm wide (one specimen measured). Carapace dark orange brown, finely veined with black. Chelicerae yellow brown, marked with dark brown. Legs orange yellow, darker at bases. Abdomen gray or black, suffused with orange;

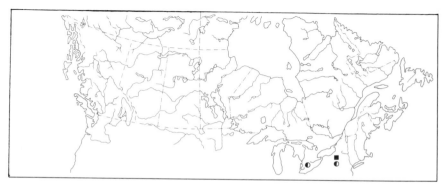

Map 47. Collection localities of *Scotinella deleta* (■) and *S. divinula* (◐).

with scutum covering most of dorsum, and with less conspicuous epigastric scutum. Femur of palpus with hooked apophysis on ventral side. Tibia of palpus slightly longer than wide, with small ventral prominence and large retrolateral apophysis less than one and one-half times as long as wide; dorsal process of retrolateral apophysis extending slightly farther distad than ventral process, thickest at base, gradually tapered to fine tip (Fig. 302). Tegulum convex, without apophysis; embolus broad at base, abruptly narrowed to longer distal part (Fig. 301).

Female.   Unknown.

Comments.   The male of *S. deleta* is distinguished from males of other species by the following combination of characters: dorsal process of retrolateral tibial apophysis extending only slightly farther distad than ventral process, thickest at base and gradually tapered to fine tip; retrolateral apophysis less than one and one-half times as long as wide; embolus with slender distal part longer than basal part. The female of *deleta* is unknown.

Range.   Known only from the type locality in northern New York.

Biology.   The habitat is unrecorded. The type male was collected in October.

Scotinella divinula (Gertsch)

Figs. 303–306; Map 47

Phrurolithus divinulus Gertsch, 1941a:6, figs. 7–9.

Male.   Total length approximately 1.75 mm; carapace 0.82 mm long, 0.69 mm wide (one specimen measured). Carapace dark orange, indistinctly veined with black, darker in eye area. Chelicerae pale orange, indistinctly marked with black. Legs orange. Abdomen dark orange, indistinctly veined and suffused with black; with large dorsal scutum. Femur of palpus with hooked apophysis on ventral side. Tibia of palpus with large retrolateral apophysis less than one and one-half times as long as wide; dorsal process of retrolateral apophysis extending as far distad as ventral process, thickest at base and gradually tapered to fine tip (Fig. 306). Tegulum convex, without apophysis; embolus broad at base, with slender distal part approximately as long as base (Fig. 303).

Female.   Total length approximately 1.75 mm; carapace 0.77 mm long, 0.62 mm wide (one specimen measured). General structure and color essentially as in male but abdomen without dorsal scutum. Epigynum with elongate plate convex along posterior margin; copulatory openings inconspicuous, located at lateral margins of small atrial depressions (Fig. 304). Copulatory tubes broad anteriorly, becoming slender posteriorly, sinuous; spermathecae ovoid, rather large (Fig. 305).

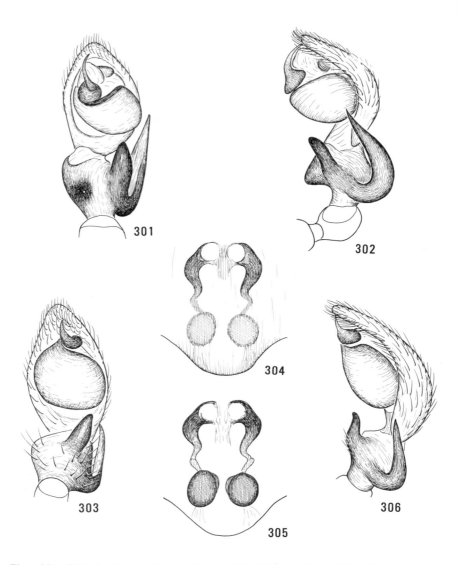

Figs. 301–306. Genitalia of *Scotinella* spp. 301, 302, *S. deleta*. 301, Palpus of male, ventral view; 302, Palpus of male, retrolateral view. 303–306, *S. divinula*. 303, Palpus of male, ventral view; 304, Epigynum; 305, Spermathecae, ventral view; 306, Palpus of male, retrolateral view.

Comments.   Specimens of *S. divinula* are distinguished from those of other species by the following combination of characters: dorsal process of retrolateral tibial apophysis in males extending as far distad as ventral process, thickest at base and gradually tapered to fine tip; retrolateral apophysis less than one and one-half times as long as wide; embolus with slender distal part as long as base; copulatory openings of females both at same level on epigynal plate; epigynum without paired dark smooth swellings; copulatory tubes sinuous; spermathecae broadly ovoid, with fingerlike spermathecal organ.

Range.   Southern Ontario to North Carolina.

Biology.   Nothing is recorded.

Family Anyphaenidae Bertkau

Spiders of the family Anyphaenidae are long-legged active hunters. Some inhabit the foliage of trees and shrubs and can be collected by beating trays or sweep nets; others mainly inhabit leaf litter or crevices under logs and stones on forest floors and must be sought by means of sifting screens, pitfall traps , or by hand searches.

The leg tarsi of the anyphaenids have relatively few long flattened setae, which are expanded and truncated at the tips and arranged in double rows as claw tufts (Platnick 1974, Platnick and Lau 1975). A second feature of these spiders is the position of the tracheal spiracle, which is located midway between the genital groove and the spinnerets or even farther anteriad. The tracheae themselves are large in cross section and extend through the pedicel into the cephalothorax and leg bases. Moreover, males have relatively larger tracheae than females, and Platnick (1974) surmises that this is correlated with the great vigor with which the males perform courtship. In the European species *Anyphaena accentuata* (Walckenaer) the males attract females by walking jerkily from side to side, bending the legs up and down, and wagging their abdomens while the palpi are set in vibration; then follows a series of audible bursts of sound produced by the extremely rapid pulsation of both the abdomen and the front legs (Bristowe 1971).

Another feature of the anyphaenids is the relative ornateness of the external male genitalia. The retrolateral tibial apophysis on the palpal tibia may be diverse in shape. The parts of the bulb itself are complex, and there may be sexual swellings and processes, as well as modified setae, on the legs.

Description.   Total length 2.90−8.25 mm. Carapace (Figs. 319, 332) ovoid in dorsal view, longer than wide, widest at level of coxae II or of coxae II and III, highest at level of dorsal groove or along middle third, with short shallow dorsal groove, with sparse covering of short pale recumbent setae, and with few long erect setae on front and in eye area; surface of carapace off-white, yellow, or orange, usually with narrow black band at lateral margins and with indistinct longitudinal lateral bands. Eyes moderately large, arranged in two transverse rows, narrowly ringed with black; anterior row slightly recurved, with median eyes smaller and slightly closer to lateral eyes than to each other; posterior row slightly longer than anterior row, straight or slightly procurved, approximately uniform in size, with median eyes round or ovoid, evenly spaced or with medians slightly closer to laterals than to each other. Chelicerae short or long, rather slender, hairy, rarely protruding; promargin of fang furrow with three to six teeth; retromargin with five to nine minute teeth. Palp-coxal lobes longer than wide, somewhat convex ventrally, with lateral margin straight (Fig. 316), slightly convex, or with angular concavity (Fig. 308). Legs (Figs. 319, 332) off-white, yellow, or orange, sometimes with dark markings, prograde, rather long, stout or slender, with sparse scopulae, and with rather sparse claw tufts composed of relatively few setae; setae expanded and truncate at tips; trochanter IV with distinct notch at tip on ventral surface; with sparse covering of short setae; femur I with two or three dorsal macrosetae, one to three prolaterals near tip of segment; coxae, femora, and tibiae II−IV in males sometimes thickened or bearing swellings, processes, or modified setae (Figs. 324, 326). Abdomen elongate, ovoid, widest at middle or in anterior third, with sparse covering of short setae and with cluster of long curved erect setae at anterior end; off-white, yellow, or orange, often with dark pattern; with tracheal spiracle located midway between genital groove and spinnerets (Fig. 315) or farther anteriad (Fig. 311), and with tracheae extending into cephalothorax; anterior spinnerets touching at bases, not more heavily sclerotized than posterior spinnerets (Figs. 311, 315). Tibia of male palpus longer than wide, with variously shaped retrolateral apophysis; tegulum prominent, often branched, with small to large median apophysis; embolus short and concealed by tegulum and median apophysis (Fig. 327), or longer, hairlike (e.g., Figs. 307, 317). Epigynum of female with large angular plate or with membranous area bounded laterally by long bandlike sclerites, with or without hood; copulatory openings usually small, inconspicuous (Figs. 310, 311, 313, 321). Copulatory tubes short, sometimes coiled or looped; spermathecae in one or in two parts, bulbous or elongate, with or without spermathecal organ (e.g., Figs. 312, 314, 322).

Comments.   Specimens belonging to species in the family Anyphaenidae are distinguished from those in other families by the following combination of characters: claw tufts composed of two rows of setae, which are expanded and truncated at tips; tracheal spiracle located midway between genital groove and spinnerets or farther anteriad, and tracheae extending into cephalothorax; external male genitalia complex; legs of adult males often with swellings, processes, or modified setae; palp-coxal lobes convex on ventral surface, with lateral margins straight, convex, or with angular indentation; anterior spinnerets touching at bases and not more sclerotized than posterior spinnerets.

Platnick (1974) revised the species of Anyphaenidae found in America north of Mexico. He treated 36 species in five genera, and estimated a world fauna of "perhaps five hundred species." Six species are represented in Canada.

164

Key to genera of Anyphaenidae

1. Tracheal spiracle located approximately midway between genital groove and spinnerets (Fig. 315). Palp-coxal lobes straight or slightly convex along lateral margins (Fig. 316). Basitarsus I with one or more prolateral and one or more retrolateral macrosetae. Embolus of male not arising on palea (e.g., Figs. 317, 320). Epigynum of female without large ovoid membranous area (e.g., Figs. 313, 321, 328) 2

   Tracheal spiracle located approximately one-fourth the distance from genital groove to spinnerets (Fig. 311). Palp-coxal lobes with angular concavities on lateral margins (Fig. 308). Basitarsus I without prolateral or retrolateral macrosetae. Embolus of male arising on large flat palea (Fig. 307). Epigynum of female with large ovoid membranous area bordered by bandlike sclerites (Fig. 310) .... *Aysha* **Keyserling** (p. 166)

2(1). Leg I approximately twice as long as body (Fig. 319). Femur I with one prolateral macroseta. Retrolateral tibial apophysis of male arising near base of segment (Fig. 317) .... *Wulfila* **O. Pickard-Cambridge** (p. 169)

   Leg I only slightly longer than body (Fig. 332). Femur I with two or three prolateral macrosetae. Retrolateral tibial apophysis of male arising at tip of segment (e.g., Figs. 323, 333)......... *Anyphaena* **Sundevall** (p. 172)

Clé des genres d'Anyphænidæ

1. Stigmate trachéal situé à peu près à mi-chemin entre la gouttière génitale et les filières (fig. 315). Lobes coxo-palpaux droits ou légèrement convexes le long des marges latérales (fig. 316). Basitarse I avec une ou plusieurs macrosetæ prolatérales et une ou plusieurs macrosetæ rétrolatérales. Embolus du mâle ne sortant pas de la palea (p. ex., fig. 317 et 320). Épigyne de la femelle sans grande zone membraneuse ovoïde (p. ex., fig. 313, 321 et 328) 2

   Stigmate trachéal situé à environ le quart de la distance séparant la gouttière génitale des filières (fig. 311). Lobes coxo-palpaux avec concavités angulaires sur les marges latérales (fig. 308). Basitarse I sans macrosetæ prolatérales ou rétrolatérales. Embolus du mâle sortant d'une grosse palea aplatie (fig. 307). Épigyne de la femelle avec grande zone membraneuse ovoïde bordée de sclérites rubanés (fig. 310) .... *Aysha* **Keyserling** (p. 166)

2(1). Patte I environ deux fois plus longue que le corps (fig. 319). Fémur I avec une macroseta prolatérale. Apophyse tibiale rétrolatérale du mâle sortant près de la base du segment (fig. 317) .... *Wulfila* **O. Pickard-Cambridge** (p. 169)

   Patte I légèrement plus longue que le corps (fig. 332). Fémur I avec deux ou trois macrosetæ prolatérales. Apophyse tibiale rétrolatérale du mâle sortant à l'extrémité du segment (p. ex., fig. 323 et 333) .... *Anyphæna* **Sundevall** (p. 172)

Genus *Aysha* Keyserling

Spiders of the genus *Aysha* are active hunters on plant foliage. They are the only Canadian anyphaenids in which the tracheal spiracle is located far anteriad near the genital groove and in which the palp-coxal lobes have angular concavities on the lateral margins. The copulatory organ of males is highly complex and differs in a number of ways from that found in representatives of the other genera.

Description. Total length approximately 6.50 mm. Carapace ovoid, longer than wide, widest at level of coxa II, highest along middle third, sparsely covered with short setae and with several long erect setae on front and in eye area, with short shallow distinct dorsal groove; surface of carapace yellow orange, with indistinct darker longitudinal bands on lateral areas. Eyes moderately large, round, narrowly ringed with black, arranged in two transverse rows; anterior row slightly recurved, with median eyes slightly smaller than lateral eyes and slightly closer to laterals than to each other; posterior row slightly longer than anterior, slightly procurved, approximately uniform in size and spacing. Chelicerae rather long, slender; promargin of fang furrow with three or four teeth; retromargin of fang furrow with seven to nine minute teeth. Palp-coxal lobes approximately twice as long as wide, rather convex ventrally, with angular concavity at middle of lateral margin (Fig. 308). Legs moderately long, slender, with rather sparse claw tufts and scopulae; orange yellow, without dark markings; trochanter IV with notch at tip on ventral surface; legs of males without swellings, processes, or modified setae; femur I with three dorsal macrosetae, two prolaterals; basitarsus I without prolateral or retrolateral macrosetae. Abdomen elongate, ovoid, rather low, widest in anterior third; off-white to gray green, sometimes spotted; with tracheal spiracle located approximately one-fourth the distance from genital groove to spinnerets (Fig. 311). Tibia of male palpus approximately twice as long as wide, with long slender retrolateral apophysis (Figs. 307, 309); embolus long, hairlike, arising on margin of flat palea (Fig. 307). Epigynum of female with large ovoid membranous area bordered laterally by paired bandlike sclerites, with small hood; copulatory openings inconspicuous (Fig. 310). Copulatory tubes long, broad; spermathecae small, bulbous (Fig. 312).

Comments. Specimens of *Aysha* spp. are distinguished from those of *Wulfila* spp. and *Anyphaena* spp. by the more anterior position of the tracheal spiracle, by the possession of angular concavities on the lateral margins of the palp-coxal lobes, by the lack of macrosetae on the prolateral and retrolateral surfaces of basitarsus I, and by the possession of a large palea in the palpus of the male and a large ovoid membranous area on the epigynal plate of the female.

Platnick (1974) estimates that 30 or more species belong in the genus *Aysha*. One species is represented in Canada.

Aysha gracilis (Hentz)

Figs. 307−312; Map 48

Clubiona gracilis Hentz, 1847:452, fig. 9 (pl. 23).
Anyphaena rubra Emerton, 1890:186, figs. 1−1*b* (pl. 6).
Aysha gracilis: Bryant, 1931:119, fig. 13 (pl. 7), fig. 26 (pl. 8); Platnick 1974:252, figs. 116, 117, 140, 143.

Male.   Total length approximately 6.60 mm; carapace 2.71−2.96 mm long, 2.09−2.37 mm wide (three specimens measured). Carapace yellow orange, darkening to orange brown at anterior end, with faintly indicated darker longitudinal lateral bands. Chelicerae dark orange brown or dark brown. Legs pale orange yellow, without dark markings. Abdomen off-white to gray green, sometimes with transverse rows of minute dull red or brown spots. Tibia of palpus approximately twice as long as wide, with long slender retrolateral apophysis that arises at tip of segment (Figs. 307, 309). Embolus hairlike, arising on prolateral margin of round flat shiny palea, extending along basal and retrolateral margins of tegulum, with tip lying in groove near tip of cymbium (Fig. 307).

Female.   Total length approximately 7.25 mm; carapace 2.81−3.14 mm long, 2.09−2.57 mm wide (seven specimens measured). General structure and color essentially as in male. Epigynum with large ovoid membranous area bordered laterally by paired bandlike sclerites, with small flat hood lying anterior to deep cavity in anterior half; copulatory openings small, inconspicuous, lying at posterior ends of narrow grooves that extend posteriad along mesal margins of lateral sclerites (Fig. 310). Copulatory tubes long, broad throughout, convoluted, forming large open loop; spermathecae rather small, bulbous (Fig. 312).

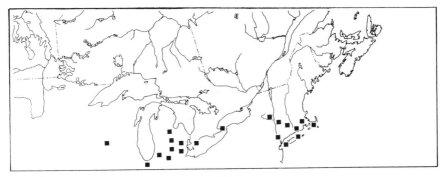

Map 48. Collection localities of *Aysha gracilis*.

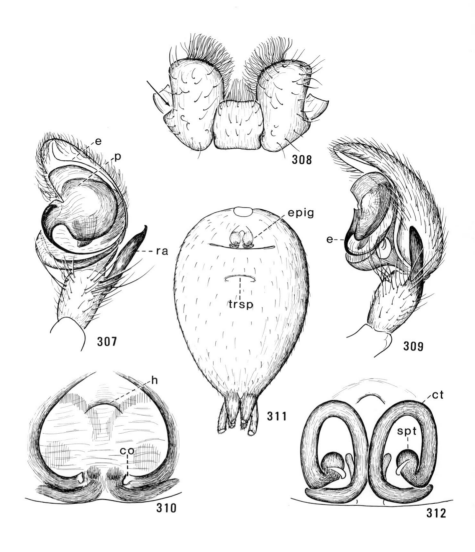

Figs. 307–312. Structures of *Aysha gracilis*. 307, Palpus of male, ventral view; 308, Palp-coxal lobes, ventral view; 309, Palpus of male, retrolateral view; 310, Epigynum; 311, Abdomen of female, ventral view; 312, Spermathecae, dorsal view. *co*, copulatory opening; *ct*, copulatory tube; *e*, embolus; *epig*, epigynum; *h*, hood; *p*, palea; *ra*, retrolateral apophysis; *spt*, spermathecae; *trsp*, tracheal spiracle.

Comments. Adults of *A. gracilis* bear a striking resemblance to those of several species of *Clubiona* (family Clubionidae) in size, color, carapace and abdominal shape, eye arrangement, chelicerae, and palp-coxal lobes but differ from these clubionids by the anterior position of the tracheal spiracle, the slenderness of the legs, the possession of two prolateral macrosetae (rather than one) on femur I, the more complex external genitalia, and the distinctive claw tufts.

Range. Texas to Florida, northward to Iowa, southern Ontario, and Massachusetts.

Biology. Specimens of *A. gracilis* have been collected from deciduous and coniferous trees and shrubs by sweep nets, and from pitcher plants, fall webworm nests, crevices in Malaise traps, and houses. Adults of both sexes have been taken in every month of the year. Kaston (1948) reports seeing an egg sac in June, attached to a leaf.

Genus *Wulfila* O. Pickard-Cambridge

Spiders of the genus *Wulfila* are active hunters on foliage or in plant litter. Little is known about their behavior, and they appear to be generally scarce in collections. Their most striking features are the conspicuous elongation of the legs, leg I being twice or more the length of the entire body, and the pale color of their bodies.

Description. Total length 2.50–4.50 mm. Carapace (Fig. 319) ovoid, longer than wide, widest at level of coxae II and III, highest at dorsal groove, with sparse covering of short recumbent setae, and with short indistinct dorsal groove; surface of carapace off-white to pale yellow, with narrow dark band along lateral margins and with dark longitudinal lateral bands formed from streaks radiating from dorsal groove area. Eyes moderately large, usually round, narrowly ringed with black, in two transverse rows; anterior row slightly recurved, with median eyes smaller than laterals and slightly closer to laterals than to each other; posterior row slightly longer than anterior row, with median eyes somewhat ovoid, similar to laterals in size, slightly closer to laterals than to each other. Chelicerae rather short, slender, off-white with indistinct darker markings; promargin of fang furrow with four to six teeth; retromargin of fang furrow with six to eight minute teeth. Palp-coxal lobes (Fig. 316) approximately twice as long as wide, convex ventrally, with lateral margin straight. Legs long, slender (particularly leg I) (Fig. 319), off-white or pale yellow, without dark markings; with thin claw tufts and sparse scopulae, and with short pale setae; trochanter IV with deep notch at tip on ventral surface; femur I with three dorsal macrosetae, one prolateral; basitarsus I without dorsal macrosetae, with three prolaterals, three retrolaterals, two pairs of ventrals; coxae and other leg segments in males without swellings, processes, or modified setae. Abdomen (Fig. 319) elongate, ovoid, widest at middle, high anteriad, with sparse covering of short pale setae and with cluster of long curved erect setae at anterior end; off-white to pale yellow, with numerous small dark spots; tracheal spiracle situated approximately midway between genital groove and spinnerets. Tibia of male palpus with stout retrolateral apophysis arising near base of segment; tegulum and median apophysis prominent; embolus long, slender, sinuous (Figs. 317, 318). Epigynum of female with large convex pale plate with spermathecae visible through cuticle; copulatory openings minute (Fig. 313). Copulatory tubes short, slender, sinuous; spermathecae each in mesal and lateral parts (Fig. 314).

Comments. Specimens of *Wulfila* spp. are distinguished from those of *Aysha* spp. and *Anyphaena* spp. by having conspicuously long and slender legs (particularly leg I) and pale bodies, and by having the retrolateral tibial apophysis on the male palpus arise near the base of the segment.

Platnick (1974) estimates a world fauna of at least 50 species in the genus *Wulfila*. One species is represented in Canada.

Wulfila saltabundus (Hentz)

Figs. 313–319; Map 49

Clubiona saltabunda Hentz, 1847:453, fig. 23 (pl. 23).
Wulfila saltabunda: Platnick, 1974:243, figs. 81, 82, 89, 99.

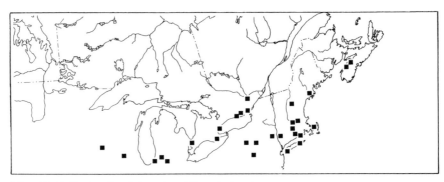

Map 49. Collection localities of *Wulfila saltabundus*.

Figs. 313–319. Structures of *Wulfila saltabundus*. 313, Epigynum; 314, Spermathecae, dorsal view; 315, Abdomen of female, ventral view; 316, Palp-coxal lobes, ventral view; 317, Palpus of male, ventral view; 318, Palpus of male, retrolateral view; 319, Body of female, dorsal view. *co*, copulatory opening; *ct*, copulatory tube; *e*, embolus; *ma*, median apophysis; *ra*, retrolateral apophysis; *spt*, spermatheca; *teg*, tegulum; *trsp*, tracheal spiracle.

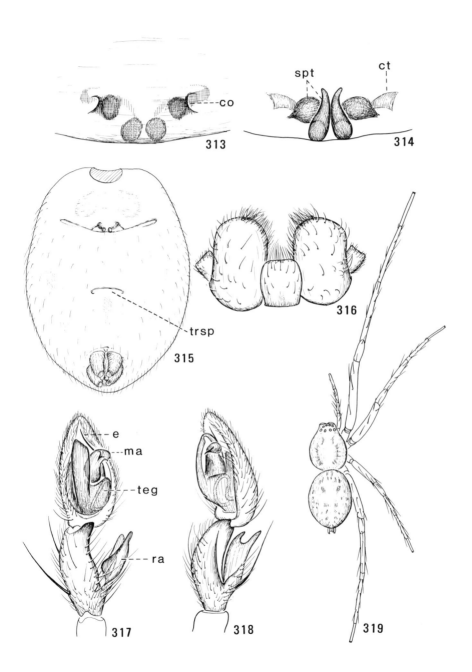

313

314

spt   ct

co

trsp

315

316

e
ma
teg
ra

317

318

319

Male. Total length approximately 3.15 mm; carapace 1.39−1.66 mm long, 1.09−1.29 mm wide (eight specimens measured). Carapace pale yellow, with narrow interrupted gray green band along lateral margin and with narrow band of gray green streaks that radiate from dorsal groove area. Chelicerae off-white with indistinct gray smudges. Legs pale yellow or orange yellow, without dark markings. Abdomen pale yellow with several curved transverse rows of small gray green spots. Patella and tibia of palpus with five to seven long stout sinuous macrosetae on prolateral and dorsal surfaces. Tibia of palpus approximately four times as long as wide, arched prolaterad, with long broad retrolateral apophysis arising near base of segment; retrolateral apophysis narrow at base, widening toward tip, excavated and toothed at tip (Figs. 317, 318). Tegulum convex, broad in basal half, extending nearly to tip of alveolus; median apophysis rather prominent, curved, with two small hard teeth at tip, attached to tegulum by membranous area near midline of bulb; embolus slender, sinuous, arising in cavity formed by tegulum and median apophysis, with tip lying on groove near tip of cymbium (Fig. 317).

Female. Total length approximately 3.60 mm; carapace 1.31−1.65 mm long, 0.97−1.25 mm wide (seven specimens measured). General structure (Figs. 315, 316, 319) and color essentially as in male. Epigynum with large pale convex rectangular plate, with spermathecae visible through cuticle covering small sloped area near posterior margin of plate; copulatory openings minute, located at level of lateral ends of spermathecae (Fig. 313). Copulatory tubes short, slender, sinuous; spermathecae each formed of two parts, with mesal part elongate, tapered, and with lateral part smaller, angular (Fig. 314).

Comments. Specimens of *W. saltabundus* are distinguished from those of *Aysha* spp. and *Anyphaena* spp. by their long slender legs and pale bodies, and by the position of the retrolateral tibial apophysis in males.

Range. Texas to Florida, northward to Nebraska, southern Ontario, and Nova Scotia.

Biology. Specimens of *W. saltabundus* have been collected from trees and shrubs by sweep nets and from leaf litter by pitfall and vacuum traps. A few were taken in the crevices of Malaise traps. Mature specimens of both sexes have been taken from April to August. Kaston (1948) reports a female in July attending egg sacs attached to the underside of a leaf.

Genus *Anyphaena* Sundevall

Spiders of the genus *Anyphaena* are wandering active hunters on plant foliage or in litter on the ground. They exhibit a great diversity of size, color, and external genitalia. Little is recorded about their behavior or their ecology.

Description. Total length 3.00−6.50 mm. Carapace (Fig. 332) ovoid, longer than wide, widest at level of leg II or of legs II and III, sometimes abruptly narrowed anteriad at level of leg I, sometimes somewhat angular at lateral margins, highest at level of dorsal groove, with sparse covering of short pale recumbent setae and with few long erect setae on front and eye area; with short pale distinct dorsal groove; surface of carapace orange or yellow orange, with dark margins and longitudinal lateral bands. Eyes moderately large, round, narrowly ringed with black, arranged in two transverse rows; anterior row slightly recurved, with median eyes smaller than others, uniformly spaced or with median eyes slightly closer to lateral eyes than to each other; posterior row slightly longer than anterior row, straight or slightly procurved, approximately uniform in size and spacing. Chelicerae rather short, slender, setaceous; yellow to orange brown; promargin of fang furrow with three or four teeth; retromargin of fang furrow with five to eight minute teeth. Palp-coxal lobes longer than wide, somewhat convex ventrally, straight or convex along lateral margins. Legs (Fig. 332) rather long, stout, with sparse claw tufts and scopulae, with sparse covering of short pale setae; yellow or orange, with or without dark markings; trochanter IV with deep notch at tip on ventral surface; coxae, femora, and tibiae II−IV of males sometimes thickened or otherwise modified with special swellings, processes, or setae (Figs. 324−326); femur I with two or three dorsal macrosetae, two or three prolaterals; basitarsus I with zero to two dorsal macrosetae, one to three prolaterals, two or three retrolaterals, and one or two pairs of ventrals. Abdomen (Fig. 332) elongate, ovoid, widest at or near middle, with sparse covering of short pale setae and with cluster of long curved erect setae at anterior end; off-white to orange, with numerous small gray or black spots; tracheal spiracle located approximately midway between genital groove and spinnerets. Tibia of male palpus longer than wide, with diverse retrolateral apophyses that arise at tip of segment (e.g., Figs. 323, 330, 339); tegulum convex, branched; median apophysis usually prominent; embolus short and concealed or long and slender (e.g., Figs. 320, 327, 331). Epigynum of female usually with angular convex plate, with or without hood; copulatory openings inconspicuous (Figs. 321, 328, 334, 337). Copulatory tubes usually short, stout, sometimes longer, coiled around spermathecae (Fig. 338); spermathecae bulbous or elongate (e.g., Figs. 322, 338).

Comments. Specimens of *Anyphaena* spp. are distinguished from those of *Aysha* spp. by the more posterior position of the tracheal spiracle, the lack of concavities on the lateral margins of the palp-coxal lobes, the possession of one or more macrosetae on prolateral and retrolateral surfaces of basitarsus I, and by the absence of a palea in the male palpus and the absence of a large membranous area in the female epigynum. Specimens of *Anyphaena* spp. are distinguished from those of *Wulfila* spp. by the stouter legs, colored bodies, the possession of more than one prolateral macroseta on basitarsus I, and the distal placing of the male retrolateral tibial apophysis.

Platnick (1974) estimates a world fauna of at least 55 species in the genus *Anyphaena*. Four species are represented in Canada.

Key to species of *Anyphaena*

1. Male 2
   Female 5
2(1). Coxae III and IV either with dense covering of short stout setae (Fig. 324) or with ventral processes (Fig. 325). Basitarsus I with one or no dorsal macrosetae 3
   Coxae III and IV with neither dense covering of short stout setae nor ventral processes. Basitarsus I with two dorsal macrosetae 4
3(2). Coxae III and IV with dense covering of short stout setae, without ventral processes (Fig. 324). Tibia III with thickened conelike ventral macrosetae (Fig. 326). Retrolateral tibial apophysis V-shaped (Fig. 323) ....
   .... *celer* (**Hentz**) (p. 175)
   Coxae III and IV without dense covering of short stout setae, with ventral processes (Fig. 325). Tibia III with unmodified ventral macrosetae. Retrolateral tibial apophysis short, broad (Fig. 330) ....
   .... *pectorosa* L. **Koch** (p. 176)
4(2). Retrolateral tibial apophysis short, blunt (Fig. 339). Embolus hidden (ventral view, Fig. 336). Basitarsus I with two pairs of ventral macrosetae ......
   .... *pacifica* (**Banks**) (p. 179)
   Retrolateral tibial apophysis boat-shaped (Fig. 333). Embolus prominent (Fig. 331). Basitarsus I with one pair of ventral macrosetae ....
   .... *aperta* (**Banks**) (p. 181)
5(1). Basitarsus I with one or no dorsal macrosetae 6
   Basitarsus I with two dorsal macrosetae 7
6(5). Epigynum with hood (Fig. 321) .... *celer* (**Hentz**) (p. 175)
   Epigynum without hood (Fig. 328) .... *pectorosa* L. **Koch** (p. 176)
7(5). Epigynum without hood (Fig. 337). Copulatory tube coiled around anterior end of spermatheca (Fig. 338). Basitarsus I with two pairs of ventral macrosetae
   .... *pacifica* (**Banks**) (p. 179)
   Epigynum with hood (Fig. 334). Copulatory tube not coiled (Fig. 335). Basitarsus I with one pair of ventral macrosetae ....
   .... *aperta* (**Banks**) (p. 181)

Clé des espèces d'*Anyphæna*

1. Mâle 2
   Femelle 5
2(1). Hanches III et IV densément couvertes de courtes soies trapues (fig. 324) ou de processus ventraux (fig. 325). Basitarse I avec une macroseta dorsale ou sans aucune 3
   Hanches III et IV non densément couvertes de soies courtes et trapues, ni de processus ventraux. Basitarse I avec deux macrosetæ dorsales 4
3(2). Hanches III et IV densément couvertes de courtes soies trapues, sans processus ventraux (fig. 324). Tibia III avec macrosetæ ventrales épaissies et coniques (fig. 326). Apophyse tibiale rétrolatérale en forme de V (fig. 323) ......
   .... *celer* (**Hentz**) (p. 175)
   Hanches III et IV non densément couvertes de soies courtes et trapues, avec processus ventraux (fig. 325). Tibia III avec macrosetæ ventrales non modifiées. Apophyse tibiale rétrolatérale courte, large (fig. 330) ....
   .... *pectorosa* L. **Koch** (p. 176)

Anyphaena celer (Hentz)

Figs. 320–324, 326; Map 50

Clubiona celer Hentz, 1847:452, fig. 20 (pl. 23).
Anyphaena incerta Keyserling, 1887:452, fig. 22 (pl. 6).
Anyphaena celer: Simon, 1897:96; Platnick 1974:214, figs. 1, 9, 10, 18.

Male.   Total length approximately 4.35 mm; carapace 2.06–2.12 mm long, 1.56–1.62 mm wide (five specimens measured). Carapace yellow orange, with narrow black line along lateral margins and with interrupted orange gray band as wide as eye area covering much of median area. Chelicerae yellow orange. Legs orange yellow, with few faint gray rings or spots; femur III thickened distad; tibia III with thickened conelike ventral macrosetae (Fig. 326); coxae III and IV densely

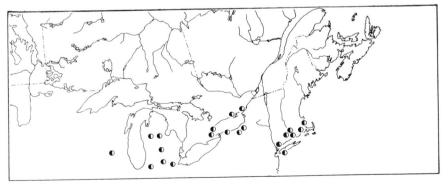

Map 50. Collection localities of *Anyphaena celer*.

covered with short stout dark setae (Fig. 324). Abdomen orange yellow, with numerous small gray smudges or spots. Tibia of palpus slightly longer than wide, with shiny prominence on ventral surface, and with stout V-shaped retrolateral apophysis; retrolateral apophysis with ventral process concave and fluted at tip, with dorsal process slightly longer than ventral process, flattened, hollowed along ventral margin, tapered to broad point (Figs. 320, 323). Tegulum branched at base, with stout median piece that extends far distad and with shorter retrolateral piece; median apophysis short, slender, curved; embolus slender, slightly curved, arising near middle of bulb, with tip lying on broad groove near tip of cymbium (Fig. 320).

Female. Total length approximately 5.35 mm; carapace 1.93, 2.07 mm long, 1.39, 1.57 mm wide (two specimens measured). General structure and color essentially as in male but legs lacking special swellings or setae. Epigynum with rectangular plate bearing hood, triangular median septum, and membranous atrium; copulatory openings inconspicuous, pocketlike, located in grooves (Fig. 321). Copulatory tubes not apparent in dorsal view; spermathecae bulbous, separated by their width or less (Fig. 322).

Comments. The V-shaped retrolateral tibial apophysis and dense covering of short stout setae on coxae III and IV distinguish males of *A. celer* from those of the other species in Canada. The females have an epigynal hood and have one or no dorsal macrosetae on basitarsus I.

Range. Texas to Florida, northward to Wisconsin and southern Ontario.

Biology. Specimens of *A. celer* have been collected from trees and shrubs by sweep nets and from leaf litter by pitfall traps and vacuum traps. Mature specimens of both sexes have been taken in all months of the year.

Anyphaena pectorosa L. Koch

Figs. 325, 327−330; Map 51

Anyphaena pectorosa L. Koch, 1866:198, figs. 131, 132 (pl. 8); Platnick 1974:230, figs. 51, 55, 59, 74, 75.
Anyphaena calcarata Emerton, 1890:187, figs. 3−3d (pl. 6).

Male. Total length approximately 4.50 mm; carapace 1.48−2.43 mm long, 1.26−1.98 mm wide (four specimens measured). Carapace yellow orange, with dark lateral margins and with indistinct longitudinal lateral bands formed by short, lighter and darker bands radiating from dorsal groove area. Chelicerae orange yellow or pale orange, lightly suffused with gray. Legs yellow (coxae to patellae) and orange (tibiae and tarsi); coxa II swollen ventrally, with cluster of short stiff setae; coxa III with hairy blunt prominence and pointed process on ventral surface; coxa IV with long pointed process on ventral side (Fig. 325). Abdomen off-white, with few paired indistinct gray smudges. Tibia of palpus approximately one and one-half times as long as wide, with small hooked ventral process and with short broad toothed retrolateral apophysis (Figs. 327, 330).

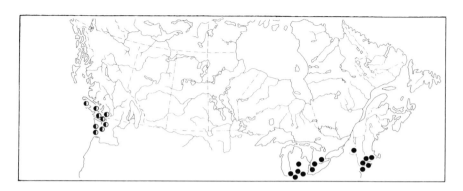

Map 51. Collection localities of *Anyphaena pectorosa* (■) and *A. aperta* (○).

Cymbium with small basal process that interlocks with retrolateral tibial apophysis (Fig. 330). Tegulum branching at base into three pieces, a slender prolateral one, a large one that bends dorsad at tip toward alveolar wall, and a short retrolateral one; median apophysis strongly hooked at tip; embolus short, hidden by tegulum (Fig. 327).

Female. Total length approximately 5.00 mm; carapace 1.48−2.41 mm long, 1.53−1.93 mm wide (six specimens measured). General structure and color essentially as in male but leg coxae without swellings, processes, or modified setae. Epigynum with broad angular convex plate, without hood; copulatory openings small, conjoined at midline near posterior margin of plate (Fig. 328). Copulatory tubes not visible in ventral or dorsal view; spermathecae bulbous, touching at midline (Fig. 329).

Comments. Males of *A. pectorosa* are distinguished from those of the other species in Canada by the ventral processes on coxae III and IV. Females can be distinguished by the presence of one or no dorsal macrosetae on basitarsus I and by the lack of a hood in the epigynum.

Range. Texas to Florida, northward to Illinois and southern Ontario.

Biology. Specimens of *A. pectorosa* have been collected from tree and shrub foliage by sweep nets, from deciduous leaf litter by pitfall traps, and from crevices beneath stones or in Malaise traps. Mature males have been taken from April to September, and mature females from April to August. Kaston (1948) found females attending egg sacs in July and August; the sacs were in a rolled dry leaf or in the fold of a grass blade that had been closed lengthwise edge to edge.

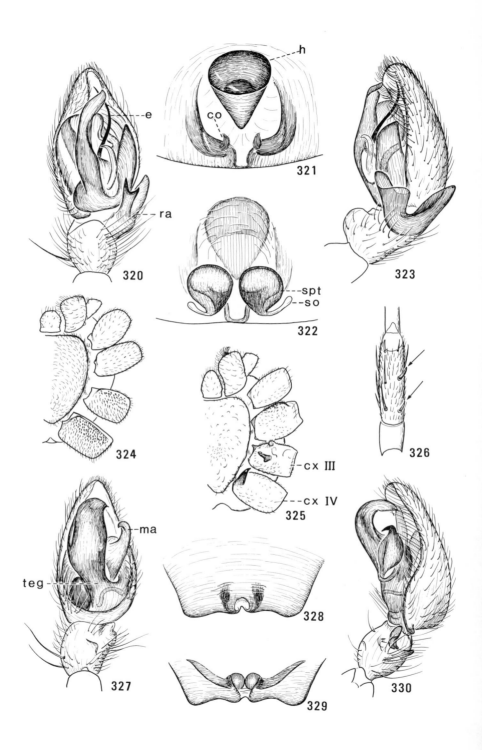

Anyphaena pacifica (Banks)

Figs. 336—339; Map 52

Gayenna pacifica Banks, 1896:63.
Anyphaena pacifica: Simon, 1897:96; Platnick 1974:236, figs. 63, 66, 68, 70.
Anyphaena mundella Chamberlin, 1919*a*:12, fig. 3 (pl. 6).
Anyphaena intermontana Chamberlin, 1920:200, fig. 22-6.
Gayenna saniuana Chamberlin & Gertsch, 1928:185.
Anyphaena pomona Chamberlin & Ivie, 1941:23, fig. 16 (pl. 2).
Gayenna jollensis Schenkel, 1950:77, fig. 27.

Male. Total length approximately 5.60 mm; carapace 2.34—2.45 mm long, 1.94—2.12 mm wide (five specimens measured). Carapace orange, with dark margins and with dark longitudinal lateral bands formed of lighter and darker bands radiating short distance from dorsal groove area. Chelicerae orange. Legs dull orange, without dark markings; basitarsus I with two dorsal macrosetae and with two pairs of ventral macrosetae. Abdomen pale orange, with pattern of small black spots and streaks, those on posterior half sometimes forming four or five short thick chevrons, sometimes purple in tone. Tibia of palpus approximately

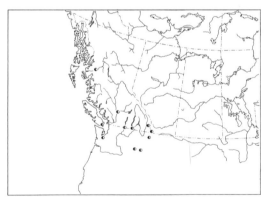

Map 52. Collection localities of *Anyphaena pacifica*.

Figs. 320—330. Structures of *Anyphaena* spp. 320—324, 326, *A. celer*. 320, Palpus of male, ventral view; 321, Epigynum; 322, Spermathecae, dorsal view; 323, Palpus of male, retrolateral view; 324, Left coxae of male, ventral view; 326, Left tibia III of male, ventral view. 325, 327—330, *A. pectorosa*. 325, Left coxae of male, ventral view. 327, Palpus of male, ventral view; 328, Epigynum; 329, Spermathecae, dorsal view; 330, Palpus of male, retrolateral view. *co*, copulatory opening; *cx*, coxa; *e*, embolus; *h*, hood; *ma*, median apophysis; *ra*, retrolateral apophysis; *so*, spermathecal organ; *spt*, spermatheca; *teg*, tegulum.

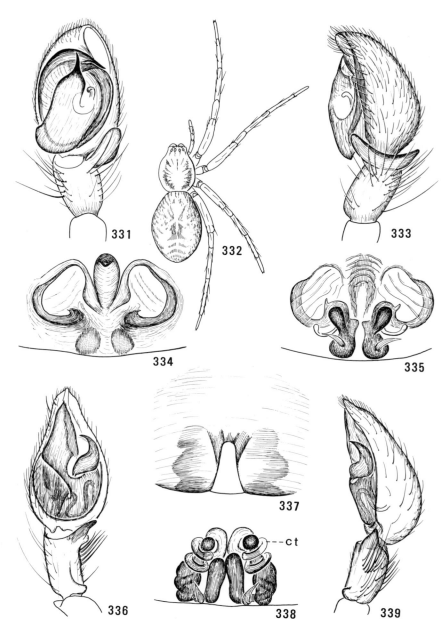

Figs. 331–339. Structures of *Anyphaena* spp. 331–335, *A. aperta.* 331, Palpus of male, ventral view; 332, Body of female, dorsal view; 333, Palpus of male, retrolateral view; 334, Epigynum; 335, Spermathecae, dorsal view. 336–339, *A. pacifica.* 336, Palpus of male, ventral view; 337, Epigynum; 338, Spermathecae, dorsal view; 339, Palpus of male, retrolateral view. *ct*, copulatory tube.

twice as long as wide, somewhat flattened on ventral side, with short blunt retrolateral apophysis and with cluster of long sinuous pale setae at base of apophysis (Figs. 336, 339). Tegulum branched at base to form large median piece and smaller piece largely hidden by median apophysis; median apophysis angular, hooked; embolus short, slender, hidden by tegulum (Fig. 336).

Female. Total length approximately 5.45 mm; carapace 2.34−2.56 mm long, 1.62−1.97 mm wide (three specimens measured). General structure and color essentially as in male. Epigynum with rectangular convex hairy plate; plate without hood, with elongate narrow atrium; copulatory openings conjoined at midline (Fig. 337). Copulatory tubes long, slender, transparent, tightly coiled around anterior ends of spermathecae; spermathecae long, twisted, extending posteriad, then mesad and anteriad (Fig. 338).

Comments. Specimens of *A. pacifica* are distinguished from those of the other species by the presence of two dorsal macrosetae and two pairs of ventral macrosetae on basitarsus I. Males are further distinguished by the short blunt retrolateral apophysis and cluster of setae on the tibial apophysis, and females are further distinguished by the coiled copulatory tubes.

Range. California to British Columbia, inland to New Mexico and Alberta.

Biology. Specimens of *A. pacifica* have been collected from leaf litter by pitfall traps and from crevices under stones and in houses. Mature males have been taken from February to July, and mature females in all months.

Anyphaena aperta (Banks)

Figs. 331−335; Map 51

Gayenna aperta Banks, 1921:100, fig. 3.
Anyphaena aperta: Bryant, 1931:114, fig. 35 (pl. 8); Platnick 1974:241, figs. 135−137.

Male. Total length approximately 4.35 mm; carapace 1.87−2.13 mm long, 1.47−1.72 mm wide (four specimens measured). Carapace dark orange, with margins black and with alternating pale and dark bands radiating from dorsal groove area. Chelicerae dark orange brown with black markings at base, paler distally. Legs with all femora (distal two-fifths), patellae, and tibiae and with basitarsus IV often nearly black and somewhat iridescent, otherwise yellow orange; basitarsus I with two dorsal macrosetae and one pair of ventrals. Abdomen off-white or pale yellow, with numerous paired black spots, some of the spots partly joining to form indistinct chevrons on posterior half. Tibia of palpus approximately twice as long as wide, with many long slender setae, and with small boat-shaped retrolateral apophysis (Fig. 333). Tegulum convex, tapered abruptly at middle; median apophysis minute, angled; embolus slender, dark, curved around prolateral margin of tegulum, with tip lying near boat-shaped cavity in margin of cymbium (Fig. 331).

Female. Total length approximately 5.50 mm; carapace 2.10−2.65 mm long, 1.68−2.05 mm wide (nine specimens measured). General structure and color (Fig. 332) essentially as in male but often somewhat paler. Epigynum with small hood and large atrium divided at midline by hood; copulatory openings concealed at posterior end of atrium (Fig. 334). Copulatory tubes short, stout; spermathecae small, nearly touching (Fig. 335).

Comments. Specimens of *A. aperta* are distinguished from those of the other species by the presence of two dorsal macrosetae and one pair of ventrals on basitarsus I, by the boat-shaped retrolateral tibial apophysis of the male, and by the presence of a hood in the epigynum of the female. Both sexes may have some of the leg segments darkened.

Range. California to British Columbia.

Biology. Specimens of *A. aperta* have been collected from redwoods, red cedars, firs, spruces, and lodgepole pines by beating trays and from oaks and deciduous shrubs by sweep nets; occasional specimens were found on tents or in the crevices of Malaise traps. Mature males have been taken from March to September and mature females from March to November.

Glossary of anatomical terms

abdomen   The posterior body division of a spider, divided from the cephalothorax by the pedicel.

alveolus   A cuplike cavity on the ventral side of the cymbium of the male palpus containing the genital bulb.

anal tubercle   A small prominence at the tip of the abdomen; the anus is situated on its ventral surface.

anteriad   Toward the anterior end of the body.

anterior   Pertaining to the foremost end of the body or of one of its main divisions.

anterolaterad   Toward the anterior end and the side.

anterolateral   Pertaining to the anterior end and the side.

anteromesad   Toward the anterior end and the midline.

anteromesal   Pertaining to the anterior end and the midline.

apophysis   A spine found on the male chelicerae, palpi, or legs and usually having a sexual function.

atrial   Pertaining to the atrium.

atrium   A cavity in the epigynal plate having the copulatory openings of the female in its floor or walls; it may be partitioned by a median septum.

basad   Toward the base, or point of attachment, of an appendage or segment.

basal   Pertaining to the base of an appendage or segment.

basitarsus   The basal subdivision of the leg tarsus.

bidentate   Having two teeth.

book lungs   The paired booklike respiratory organs on the venter of the abdomen.

carapace   The dorsal plate of the cephalothorax that bears the eyes and the dorsal groove; it represents the fused terga of the cephalothoracic segments.

cephalothorax   The undivided head−thorax, or anterior body division, to which are appended the chelicerae, palpi, and legs.

chelicerae   The paired, seizing, and pinching organs attached at the anterior end of the cephalothorax; each comprises a large basal segment and a movable fang with, internally, the associated venom gland and muscles. They arise between the mouth and the palpi in the early embryo, but move anterior to the mouth and rostrum during embryonic development.

clavate   Club-shaped.

claw   A short, curved, usually toothed process at the tip of the pretarsus of a leg or palpus.

claw tuft   A bundle of stiff setae at the tip of the leg tarsus. They are believed to provide adhesion on slippery surfaces.

conductor   A structure in the male palpus on which the tip of the embolus rests.

copulatory openings   The paired openings in the epigynal plate receiving the male emboli during copulation.

copulatory tubes   The paired tubes leading inward from the copulatory openings of the female and receiving the embolus of the male in copulation.

coxa (pl., coxae)   The first or most basal segment of a leg or palpus.

cuticle   The outer layer of the integument, or body wall.

cymbium   The tarsus of the male palpus, containing the alveolus on its ventral side.

dentate   Toothed.

distad   Toward the distal end of a leg or palpus.

distal   Pertaining to the end of a leg or palpus farthest from the base.

distitarsus   The distal subdivision of the leg tarsus.

distomesad   Toward the tip and the midline.

distomesal   Pertaining to the tip and the midline.

dorsad   Toward the dorsum.

dorsal   Pertaining to the uppermost surface of the body or of an appendage.

dorsal groove   A median furrow, or groove, on the carapace marking the presence of an ingrowth of the body wall on which the dilator muscles of the sucking pump are attached.

dorsum   The entire upper surface of the body; also used for the upper surface of the abdomen alone.

emarginate   Having a notched margin.

embolus   The intromittent, or inserting, organ of the male palpus.

epigastric   Pertaining to the ventral side of the abdomen; e.g., the epigastric scutum, a plate found on the abdominal venter anterior to the genital groove in some sac spiders.

epigynum   The copulatory organ of the female located in the midline anterior to the genital groove; usually with a well-sclerotized plate in which the copulatory openings are found.

fang   The piercing distal segment of the chelicera.

fang furrow   A depression along the distomesal surface of the chelicera; it receives the folded fang.

femur (pl., femora)   The third from the base and usually longest segment of a leg or palpus.

fertilization tubes   The paired tubes by which semen stored in the spermathecae of the female is conveyed to the eggs as they pass out of the body.

front   That part of the carapace between the anterior margin and the anterior row of eyes.

genital bulb   The copulatory apparatus lying within the alveolus of the cymbium on the male palpus.

genital groove   A transverse groove on the venter of the abdomen in which lie the openings of the internal genitalia (ovaries, testes) and a pair of book lungs.

hood   A pocketlike structure at the anterior end of the epigynum of some sac spiders.

integument   The body wall.

labium   The lower lip, which closes the preoral cavity behind; it develops from the sternum of the embryonic palpal segment.
laterad   Toward one side.
lateral   Pertaining to the side.
longitudinal   Lying parallel to the midline of the body or of a leg or palpus.

macroseta   An erectile seta that arises from a minute membranous area on the legs and palpi.
median   Pertaining to the middle.
median apophysis   The appendage of the tegulum on the genital bulb of the male palpus.
median septum   A raised longitudinal piece on the floor of the atrium of the epigynum.
mesad   Toward the midline.
mesal   Pertaining to the midline.
midline   An imaginary line dividing the body into right and left halves.

palea   A plate at the distal end of the genital bulb of some male sac spiders.
palp-coxal lobes   The paired mesal lobes on the prolateral surfaces of the palpal coxae.
palpus (pl., palpi)   One of a pair of leglike appendages arising between the mouth and the first pair of legs; in adult male spiders, modified as a semen-storing and copulatory organ.
patella   The fourth segment from the base on the leg or palpus; it forms a rigid piece with the tibia.
pedicel   The slender flexible connection between cephalothorax and abdomen.
posteriad   Toward the posterior end.
posterior   Pertaining to the hindmost end of the body or of one of its main divisions.
posteromesad   Toward the posterior end and the midline.
posteromesal   Pertaining to the posterior end and the midline.
posterolaterad   Toward the posterior end and the side.
posterolateral   Pertaining to the posterior end and the side.
preoral cavity   The entrance passage to the mouth.
pretarsus   The seventh or terminal segment of a leg or palpus; it bears the claws.
procurved   Denotes the anterior displacement of the ends of a transverse, otherwise straight, row (of eyes, for example) or groove.
prograde   Denotes the normal orientation of the legs in spiders with the limbs not rotated on their bases; also used to describe the mode of locomotion of such spiders.
prolaterad   Toward the prolateral surface.
prolateral   Pertaining to the lateral surface of a leg or palpus nearest the anterior end of the body when the appendage is extended at right angles to the midline.
prolaterobasad   Toward the prolateral side and the base.
prolaterobasal   Pertaining to the prolateral side and the base.

prolaterodistad   Toward the prolateral side and the end of a leg or palpus.
prolaterodistal   Pertaining to the prolateral side and the end.
prolateromesad   Toward the prolateral side and the midline.
prolateromesal   Pertaining to the prolateral side and the midline.
prolateroventrad   Toward the prolateral side and the venter.
prolateroventral   Pertaining to the prolateral side and the venter.
promargin   The anterior margin of the cheliceral fang furrow.

recurved   Denotes the posterior displacement of the ends of a transverse, otherwise straight, row (of eyes, for example) or groove.
retrolaterad   Toward the retrolateral surface.
retrolateral   The lateral surface of a leg or palpus nearest the posterior end of the body when the appendage is extended at right angles to the midline.
retrolaterobasad   Toward the retrolateral side and the base.
retrolaterobasal   Pertaining to the retrolateral side and the base.
retrolaterodistad   Toward the retrolateral side and the tip.
retrolaterodistal   Pertaining to the retrolateral side and the tip.
retromargin   The posterior margin of the cheliceral fang furrow.

scape   A median unpaired process of the epigynal plate, free at one end and usually directed posteriad.
sclerite   A thickened well-sclerotized plate in the body wall.
sclerotized   Hardened and darkened through the tanning of proteins.
scopula (pl., scopulae)   A brush of stiff, flattened setae along the ventral surface of the leg; a scopulate leg is one that is provided with a scopula.
scutum (pl., scuta)   A sclerite covering part of the abdomen; e.g., dorsal scutum, epigastric scutum.
segment   One of a series of ringlike divisions into which the body or an appendage is divided.
seminal duct   A tube that conducts semen from the interior of the male palpus to the embolus.
serrated   Notched like a saw.
seta (pl., setae)   An outgrowth of the integument secreted by a single cell and supplied with a nerve; setae form the usual covering of the spider's body and may be modified in shape, e.g., flattened or clublike.
spermatheca (pl., spermathecae)   One of a pair of semen-storing organs of the female.
spermathecal organ   A small prominence associated with the spermatheca and usually arising at or near the junction of the copulatory tube and spermatheca.
spine   A fixed, usually pointed, outgrowth of the body wall.
spinnerets   The paired appendages at the posterior end of the abdomen through which liquid silk passes from the silk glands to the outside; in three pairs: anterior, median, and posterior.
spiracle   Tracheal opening in the body wall, located on the venter of the abdomen.
sternum   The ventral wall of a body segment; also used for the fused sterna of the cephalothorax.

subtegulum   A ringlike sclerite in the wall of the genital bulb of the male palpus.

tarsus (pl., tarsi)   The sixth segment of a leg or palpus from the base; in legs, subdivided into basitarsus and distitarsus.

tegulum   A sclerite of the genital bulb in the male palpus.

tibia (pl., tibiae)   The fifth segment of a leg or palpus from the base; it forms a rigid piece with the patella.

tooth   A spine found on the chelicerae and assisting in feeding; also, a small outgrowth on the paired claws of the leg.

trachea   The internal system of tubes through which air exchange takes place, thus supplementing the book lungs; their openings are the spiracles.

transverse   Lying at right angles to the midline.

trochanter   The second segment of a leg or palpus from the base.

truncate   Squared, rather than rounded or pointed, at the tip.

tubercle   A small, fixed, usually rounded prominence in the body wall.

venom gland   The venom-secreting gland within (and sometimes beyond) the chelicera. Its duct opens on the tip of the fang.

venter   The undersurface of the body; also used for the undersurface of the abdomen or of an appendage alone.

ventrad   Toward the venter.

ventral   Pertaining to the venter.

References

Banks, N. 1891. Notes on some spiders described by Hentz. Ent. News 2:84-87.

Banks, N. 1892. The spider fauna of the Upper Cayuga Lake Basin. Proc. Acad. nat. Sci. Philad. pp. 11-81.

Banks, N. 1895. A list of spiders of Long Island, with description of new species. J. N.Y. ent. Soc. 3:76-93.

Banks, N. 1896. New North American spiders and mites. Trans. Am. ent. Soc. 23:57-77.

Banks, N. 1897. Descriptions of new spiders. Can. Ent. 29:193-197.

Banks, N. 1898. Arachnida from Baja California, and other parts of Mexico. Proc. Calif. Acad. Sci. Ser. 3. 1:205-308.

Banks, N. 1901. Some Arachnida from New Mexico. Proc. Acad. nat. Sci. Philad. 53:568-597.

Banks, N. 1904. Some Arachnida from California. Proc. Calif. Acad. Sci. Ser. 3. 3:331-376.

Banks, N. 1907. A preliminary list of the Arachnida of Indiana, with keys to families and genera of spiders. Rep. Indiana Dep. Geol. nat. Resour. 31:715-747.

Banks, N. 1910. Catalogue of nearctic spiders. Bull. U.S. natn. Mus. 72:1-80.

Banks, N. 1921. New Californian spiders. Proc. Calif. Acad. Sci. 11:99-102.

Barrows, W.M. 1940. New and rare spiders from the Great Smoky Mountain National Park region. Ohio J. Sci. 40:130-138.

Barrows, W.M., and Ivie, W. 1942. Some new spiders from Ohio. Ohio J. Sci. 42:20-23.

Blackwall, J. 1861. A history of the spiders of Great Britain and Ireland. Ray Society, London, England. 174 pp.

Blackwall, J. 1862. Description of newly discovered spiders captured in Rio de Janeiro by John Gray and the Rev. Hamlet Clark. Ann. Mag. nat. Hist. Ser. 3. 10:348-360, 421-439.

Bonnet, P. 1956. Bibliographia araneorum. Vol. 2. Imprimerie Douladoure, Toulouse. pp. 1-5058.

Bristowe, W.S. 1971. The world of spiders. 2nd ed. Collins, London. 304 pp.

Bryant, E.B. 1931. Notes on North American Anyphaeninae in the Museum of Comparative Zoology. Psyche (Camb. Mass.) 38:102-126.

Bryant, E.B. 1936. New species of southern spiders. Psyche (Camb. Mass.) 43:87-100.

Chamberlin, R.V. 1919a. New Californian spiders. J. Ent. Zool. 12:1-17.

Chamberlin, R.V. 1919b. New western spiders. Ann. ent. Soc. Am. 12:239-260.

Chamberlin, R.V. 1920. New spiders from Utah. Can. Ent. 52:193-201.

Chamberlin, R.V. 1925. Diagnoses of new American Arachnida. Bull. Mus. comp. Zool. Harv. Univ. 67:211-248.

Chamberlin, R.V., and Gertsch, W.J. 1928. Notes on spiders from southeastern Utah. Proc. biol. Soc. Wash. 41:175-188.

Chamberlin, R.V., and Gertsch, W.J. 1930. On fifteen new North American spiders. Proc. biol. Soc. Wash. 43:137-144.

Chamberlin, R.V., and Ivie, W. 1933. Spiders of the Raft River Mountains of Utah. Bull. Univ. Utah biol. Ser. 23(4). 53 pp.

Chamberlin, R.V., and Ivie, W. 1935. Miscellaneous new American spiders. Bull. Univ. Utah biol. Ser. 26(4). 79 pp.

Chamberlin, R.V., and Ivie, W. 1941. Spiders collected by L.W. Saylor and others, mostly in California. Bull. Univ. Utah biol. Ser. 31(8). 49 pp.

Chamberlin, R.V., and Ivie, W. 1946. On several new American spiders. Bull. Univ. Utah biol. Ser. 36(13). 15 pp.

Clerck, C. 1757. Aranei suecici .... Stockholm, Sweden. 154 pp.

Dondale, C.D., and Redner, J.H. 1976. A rearrangement of the North American species of *Clubiona*, with descriptions of two new species (Araneida: Clubionidae). Can. Ent. 108:1155-1165.

Dondale, C.D., and Redner, J.H. 1978. The insects and arachnids of Canada. Part 5. The crab spiders of Canada and Alaska. Agric. Can. Publ. No. 1663. 255 pp.

Dondale, C.D., and Redner, J.H. 1979. Designation of a lectotype for *Phrurotimpus minutus* (Araneae: Clubionidae). J. Arach. 7:266-267.

Dondale, C.D., Redner, J.H., and Semple, R.B. 1972. Diel activity periodicities in meadow arthropods. Can. J. Zool. 50:1155-1163.

Edwards, R.J. 1958. The spider subfamily Clubioninae of the United States, Canada and Alaska (Araneae: Clubionidae). Bull. Mus. comp. Zool. Harv. Univ. 118:365-436.

Emerton, J.H. 1890. New England spiders of the families Drassidae, Agalenidae and Dysderidae. Trans. Conn. Acad. Arts Sci. 8:166-206.

Emerton, J.H. 1894. Canadian spiders. Trans. Conn. Acad. Arts Sci. 9:400-429.

Emerton, J.H. 1909. Supplement to the New England spiders. Trans. Conn. Acad. Arts Sci. 14:171-236.

Emerton, J.H. 1911. New spiders from New England. Trans. Conn. Acad. Arts Sci. 16:385-407.

Emerton, J.H. 1913. New England spiders identified since 1910. Trans. Conn. Acad. Arts Sci. 18:209-224.

Emerton, J.H. 1915. Canadian spiders, II. Trans. Conn. Acad. Arts Sci. 20:145-160.

Emerton, J.H. 1919. New spiders from Canada and the adjoining States, No. 2. Can. Ent. 51:105-108.

Emerton, J.H. 1924. New spiders from southern New England. Psyche (Camb. Mass.) 31:140-145.

Fox, I. 1938. Notes on North American spiders of the families Gnaphosidae, Anyphaenidae and Clubionidae. Iowa St. Coll. J. Sci. 12:227-243.

Gerhardt, U. 1924. Neue Studien zur Sexualbiologie und zur Bedeutung des sexuellen Grössendimorphismus der Spinnen. Z. Morph. Ökol. Tiere 1:507-538.

Gertsch, W.J. 1933. Diagnoses of new American spiders. Am. Mus. Novit. No. 637. 14 pp.

Gertsch, W.J. 1935. New American spiders with notes on other species. Am. Mus. Novit. No. 805. 24 pp.

Gertsch, W.J. 1941a. New American spiders of the family Clubionidae. I. Am. Mus. Novit. No. 1147. 20 pp.

Gertsch, W.J. 1941b. New American spiders of the family Clubionidae. II. Am. Mus. Novit. No. 1148. 18 pp.

Gertsch, W.J. 1942. New American spiders of the family Clubionidae. III. Am. Mus. Novit. No. 1195. 18 pp.

Gertsch, W.J. 1979. American spiders. 2nd ed. Van Nostrand Reinhold Company, New York. 274 pp.

Gorham, R.J., and Rheney, T.B. 1968. Envenomation by the spiders Chiracanthium inclusum and Argiope aurantia. J. Am. med. Ass. 206:1958-1962.

Hackman, W. 1954. The spiders of Newfoundland. Acta zool. fenn. 79:1-99.

Hentz, N.M. 1847. Descriptions and figures of the araneides of the United States. J. Boston Soc. nat. Hist. 5:444-478.

Holm, Å. 1960. On a collection of spiders from Alaska. Zool. Bidr. Upps. 33:109-134.

Kaston, B.J. 1938a. New spiders from New England with notes on other species. Bull. Brooklyn ent. Soc. 33:173-191.

Kaston, B.J. 1938b. North American spiders of the genus Agroeca. Am. Midl. Nat. 20:562-570.

Kaston, B.J. 1938c. Check-list of the spiders of Connecticut. Bull. Conn. St. geol. nat. Hist. Surv. 60. pp. 175-221.

Kaston, B.J. 1945. New spiders in the group Dionycha with notes on other species. Am. Mus. Novit. No. 1290. 25 pp.

Kaston, B.J. 1948. Spiders of Connecticut. Bull Conn. St. geol. nat. Hist. Surv. 70. 874 pp.

Kaston, B.J. 1972. How to know the spiders. 2nd ed. Wm. C. Brown Company Publishers, Dubuque, Iowa. 289 pp.

Kaston, B.J. 1978. How to know the spiders. 3rd ed. Wm. C. Brown Company Publishers, Dubuque, Iowa. 272 pp.

Keyserling, E. 1887. Neue Spinnen aus Amerika. VII. Verh. zool.-bot. Ges. Wien 37:421-490.

Koch, C.L. 1837. Übersicht des Arachnidensystems. Heft 1. Nürnberg, Germany. 39 pp.

Koch, C.L. 1842. Die Arachniden. Band 9. Nürnberg, Germany. 108 pp.

Koch, C.L. 1843. Die Arachniden. Band 10. Nürnberg, Germany. 142 pp.

Koch, L. 1864. Die europäischen Arten der Arachnidengattung *Cheiracanthium*. Abh. naturh. Ges. Nürnberg (1864):137-162.

Koch, L. 1866. Die Arachniden-Familie der Drassiden. Hefte 1-6. Nürnberg, Germany. 304 pp.

Levi, H.W. 1951. New and rare spiders from Wisconsin and adjacent states. Am. Mus. Novit. No. 1501. 41 pp.

Lessert, R. de 1905. Arachniden Graubündens. Revue suisse Zool. 13:621-661.

Linnaeus, C. 1758. Systema naturae, . . . 10th ed. Tome 1. Stockholm, Sweden. 821 pp.

Locket, G.H., and Millidge, A.F. 1951. British spiders. Vol. I. Ray Society, London, England. 310 pp.

Mansour, F., Rosen, D., and Shulov, A. 1980*a*. Biology of the spider *Chiracanthium mildei* (Arachnida: Clubionidae). Entomophaga 25:237-248.

Mansour, F., Rosen, D., and Shulov, A. 1980*b*. Functional response of the spider *Chiracanthium mildei* (Arachnida: Clubionidae) to prey density. Entomophaga 25:313-316.

Montgomery, T.H., Jr. 1909. Further studies on the activities of araneads, II. Proc. Acad. nat. Sci. Philad. 61:548-569.

Pavesi, P. 1864. Arachnidi, *in* Notizie naturali e chimico-agronomiche sulla provinzia di Pavia. Pavia, Italy.

Peck, W.B. 1975. *Chiracanthium* in the Western Hemisphere. Proc. 6th int. Congr. Arach., Amsterdam, The Netherlands. 1974, pp. 204-209.

Peck, W.B., and Whitcomb, W.H. 1970. Studies on the biology of a spider, *Chiracanthium inclusum* (Hentz). Bull. Univ. Arkansas Agric. Exp. Stn. 753. 76 pp.

Penniman, A.J. 1978. Taxonomic and natural history notes on *Phrurolithus fratrellus* Gertsch (Araneae: Clubionidae). J. Arach. 6:125-132.

Petrunkevitch, A. 1910. Some new or little known American spiders. Ann. N.Y. Acad. Sci. 19:205-224.

Petrunkevitch, A. 1911. A synonymic index-catalogue of spiders of North, Central and South America with all adjacent islands, Greenland, Bermuda, West Indies, Terra del Feugo, Galapagos, etc. Bull. Am. Mus. nat. Hist. 29:1-791.

Pickard-Cambridge, O. 1862. List of new and rare spiders captured in 1861; being a supplement to the lists in Zool. 6493, 6862, 7553. Zoologist 20:7945-7951.

Pickard-Cambridge, O. 1874. On some new species of Drassides. Proc. zool. Soc. Lond. pp. 370-419.

Pickard-Cambridge, O. 1897. Arachnida: Araneida. Biologia cent.-am. Zool. 1:225-232.

Platnick, N.I. 1974. The spider family Anyphaenidae in America north of Mexico. Bull. Mus. comp. Zool. Harv. Univ. 146:205-266.

Platnick, N.I., and Lau, A. 1975. A revision of the *celer* group of the spider genus *Anyphaena* (Araneae: Anyphaenidae) in Mexico and Central America. Am. Mus. Novit. No. 2575. 36 pp.

Platnick, N.I., and Shadab, M.U. 1974*a*. A revision of the *tranquillus* and *speciosus* groups of the spider genus *Trachelas* (Araneae, Clubionidae) in North and Central America. Am. Mus. Novit. No. 2553. 34 pp.

Platnick, N.I., and Shadab, M.U. 1974*b*. A revision of the *bispinosus* and *bicolor* groups of the spider genus *Trachelas* (Araneae, Clubionidae) in North and Central America and the West Indies. Am. Mus. Novit. No. 2560. 34 pp.

Reiskind, J. 1969. The spider subfamily Castianeirinae of North and Central America (Araneae, Clubionidae). Bull. Mus. comp. Zool. Harv. Univ. 138:162-325.

Roddy, L.R. 1957. Some spiders from southeastern Louisiana. Trans. Am. microsc. Soc. 75:285-295.

Roddy, L.R. 1966. New species, records, of clubionid spiders. Trans. Am. microsc. Soc. 85:399-407.

Roddy, L.R. 1973. American spiders of the *Clubiona canadensis* group (Araneae: Clubionidae). Trans. Am. microsc. Soc. 92:143-147.

Schenkel, E. 1950. Spinnentiere aus dem westlichen Nordamerika, gesammelt von Dr. Hans Schenkel-Rudin. Verh. naturf. Ges. Basel 61:29-92.

Simon, E. 1897. Histoire naturelle des Araignées. Tome 2, fasc. 1. Paris, France.

Spielman, A., and Levi, H.W. 1970. Probable envenomation by *Chiracanthium mildei*, a spider found in houses. Am. J. trop. Med. Hyg. 19:729-732.

Strand, E. 1900. Zur Kenntniss der Arachniden Norwegens. K. norske Vidensk. Selsk. Forh. pp. 1-46.

Thorell, T. 1856. Recensio critica Aranearum Suecicarum, quas descripserunt Clerckius, Linnaeus, de Geerus. Nova Acta R. Soc. Sci. upsal. Ser. 3. 2:61-176.

Tullgren, A. 1946. Svensk spindelfauna. 3. Egentliga spindlar. Araneae, Fam. 5-7. Clubionidae, Zoridae och Gnaphosidae. Stockholm, Sweden. 141 pp.

Walckenaer, C.A. 1802. Faune parisienne. Insectes, ou histoire abrégée des insectes des environs de Paris. 2 vols. Paris, France. pp. 187-250.

Westring, N. 1851. Förteckning öfver de till närvarande tid Kände, i Sverige förekommande Spindlarter, utgörande ett antal af 253, deraf 132 äro nya för svenska faunan. Göteborgs K. Vetensk.-o. VitterhSamh. Handl. Ser. B. 2:25-62.

Wiehle, H. 1965. Die *Clubiona*-Arten Deutschlands, ihre natürliche Gruppierung und die Einheitlichkeit im Bau ihre Vulva. Senckenberg. biol. 46:471-505.

Worley, L.G. 1932. The spiders of Washington, with special reference to those of the San Juan Islands. Univ. Wash. Publs Biol. 1:1-63.

Index to names of families, genera, and species

(Page numbers of principal entries are in boldface;
synonyms are in italic type)